CAMBRIDGE LIBRARY COLLECTION
Books of enduring scholarly value

Darwin

Two hundred years after his birth and 150 years after the publication of 'On the Origin of Species', Charles Darwin and his theories are still the focus of worldwide attention. This series offers not only works by Darwin, but also the writings of his mentors in Cambridge and elsewhere, and a survey of the impassioned scientific, philosophical and theological debates sparked by his 'dangerous idea'.

The Teaching of Science in Cambridge

This volume contains five pamphlets which illustrate the world in which Charles Darwin moved in Cambridge, and the slow development of life and earth sciences as subjects of academic study. (Darwin himself was officially following a course of study which would fit him to become an Anglican parson.) The first pamphlet (from 1821) is a proposed series of lectures on geology by Adam Sedgwick, who taught Darwin the rudiments of the subject during a tour of north Wales. The next two are botany courses proposed by John Stevens Henslow, the mentor and close friend who first suggested that Darwin should go as naturalist on the Beagle voyage. Henslow read extracts of Darwin's letters to him to a meeting of the Cambridge Philosophical Society and published them at his own expense (the fourth pamphlet). The final pamphlet is an impassioned plea from Henslow for support for a new University Botanic Garden.

Cambridge University Press has long been a pioneer in the reissuing of out-of-print titles from its own backlist, producing digital reprints of books that are still sought after by scholars and students but could not be reprinted economically using traditional technology. The Cambridge Library Collection extends this activity to a wider range of books which are still of importance to researchers and professionals, either for the source material they contain, or as landmarks in the history of their academic discipline.

Drawing from the world-renowned collections in the Cambridge University Library, and guided by the advice of experts in each subject area, Cambridge University Press is using state-of-the-art scanning machines in its own Printing House to capture the content of each book selected for inclusion. The files are processed to give a consistently clear, crisp image, and the books finished to the high quality standard for which the Press is recognised around the world. The latest print-on-demand technology ensures that the books will remain available indefinitely, and that orders for single or multiple copies can quickly be supplied.

The Cambridge Library Collection will bring back to life books of enduring scholarly value (including out-of-copyright works originally issued by other publishers) across a wide range of disciplines in the humanities and social sciences and in science and technology.

The Teaching of Science in Cambridge

Sedgwick, Henslow, Darwin

JOHN STEVENS HENSLOW
ADAM SEDGWICK
CHARLES DARWIN

CAMBRIDGE
UNIVERSITY PRESS

CAMBRIDGE UNIVERSITY PRESS

Cambridge, New York, Melbourne, Madrid, Cape Town, Singapore,
São Paolo, Delhi, Dubai, Tokyo

Published in the United States of America by Cambridge University Press, New York

www.cambridge.org
Information on this title: www.cambridge.org/9781108002004

© in this compilation Cambridge University Press 2009

This edition first published 1846
This digitally printed version 2009

ISBN 978-1-108-00200-4 Paperback

A

SYLLABUS

OF

A COURSE OF LECTURES

ON

GEOLOGY.

BY THE

Rev. A. SEDGWICK, M.A. F.R.S. &c.

WOODWARDIAN PROFESSOR

OF THE UNIVERSITY OF CAMBRIDGE.

SECOND EDITION.

CAMBRIDGE:

Printed by James Hodson, Trinity Street.

1832.

SYLLABUS.

PART I.

CHAP. 1.—*Introduction.*

1. Great subdivisions of the materials composing the surface of the earth—organized—unorganized.

2. Further subdivisions——separation of the various branches of Natural History——distinguished from Natural Philosophy.

3. Geology—its objects, and place among the physical sciences.

4. Modes of considering the mineral kingdom—separation of Geology and Mineralogy.

5. Recent discoveries in Geology—changes in the physical surface of the earth and in the animal kingdom—connection established between geology and other branches of Natural History.

6. Ancient speculations on the Theory of the Earth false and defective—" Sacred Theories " in

B

the same manner false and defective—Examples derived from the writings of Burnet, Woodward, &c.

7. True mode of conducting Geological Speculations.

The great physical regions of the earth subdivided into River Basins. Examples. 8. Ramifications of the rivers in contiguous basins generally distinct—in some instances inosculate—remarkable examples. 9. Subdivisions of River Basins—rare examples of bifurcations in the higher part of River channels. 10. Basins of rivers generally bounded by mountain chains—Exceptions—Examples of rivers which cut through mountain chains—of rivers unconnected with great elevations, &c. General illustration of the drainage of England — Conclusions.

CHAP. III.

On the great agents by which the earth's surface is modified; and on the effects which have been produced by them during known periods.

§ 1. *Modern alluvial formations.*

1. Ascent and descent of vapours—atmospheric movements—great precipitation of water in the high regions of the earth—consequent formation of rivers, lakes, &c. and degradation of the land.

2. Formation of springs. Various modifications of them. Examples derived from the neighbourhood of Cambridge, &c.

3. Plane of Congelation—effects produced by its oscillations in Alpine regions—avalanches, glaciers, &c.

4. Degradation produced by mountain torrents —by the bursting of mountain lakes. Parallel roads of Glen Roy.—Floods of the Drave, &c.

5. Progress of rivers through mid regions— rare bifurcations.

6. *Reproductive power of rivers* in their descent to low regions—frequent bifurcations—formation of deltas and various causes of their increase and diminution. Examples from the deltas of the Rhine, the Nile, the Ganges, and the Mississippi.

7. Origin of turf bogs—mode of increase— submarine forests—floating islands, &c.

8. History of the Drainage of the Bedford Level—Practical conclusions.

9. Degradation of the solid parts of the earth by the action of the sea—Examples derived from the English Coast.

10. Great marine currents—Equatorial current —distribution of incoherent materials at the bottom of the sea. Soundings of the German ocean, &c. &c.

11. Changes produced by the operation of marine animals—coral reefs, &c. Examples.

12. Sand floods on the sea coast—formation of downs, &c.—Sand floods of desert countries.

13. *Chemical action of the elements*—decomposition of rocks—chemical deposits—Examples.
14. Time during which the several physical agents have been at work—measured by the effects produced —Difficulties in the application of natural chronometers—Examples—Conclusion.
15. Remarkable cases of organic remains in alluvial formations.

§ 2.

Products and effects of volcanic action.

1. *Earthquakes*—great extent to which they are propagated—indications of deep-seated volcanic action—precede volcanic eruptions, &c.—Examples:—Monte-Nuovo (1538); Lima (1746); Jorullo(1756); Iceland(1783); Riobamba(1797); St. Vincent (1812), &c. &c.
2. Remarkable effects of earthquakes—elevation and depression of sea coasts, &c.
3. *Volcanos*—great extent of their action—Examples from the works of Humboldt, &c.
4. Their number—linear arrangement—geographical distribution, &c.
5. Volcanic craters—composition and form—relations to other formations—*craters* of *Elevation*.
6. *Eruptions.* (1). Vapours—(2). Ashes; sand; pumice; volcanic bombs; fragments of lava; of

8

rocks, sometimes unaltered, &c.—(3). Streams of melted lava from the lips or side of the crater.

7. Currents of lava—their fluidity--heat—velocity—modification in passing from a fluid to a solid state, extent, &c.—Examples.

8. Streams of mud not originating in the crater—their origin.

9. Eruptions of mud, &c.—Examples—sometimes accompanied with bituminous matter, fish, &c.

10. Greatest eruptions after long intervals of repose. Historical view of the eruptions of Vesuvius, &c.

11. Degradation of craters—by atmospheric action—by earthquakes—by the slow chemical action of volcanic vapours, &c.

12. *Structure of lavas*—scoriaceous vesicular —glassy—stony—Experiments of Watts—Basaltic and trachytic lava—Obsidian—Pearl-stone—Pumice, &c.

13. Volcanic minerals—their origin—great number of species ;
    (1.) Sublimated in volcanic vapours and deposited in crevices ;
    (2.) Crystallized from a state of igneous fusion ;
    (3.) Formed by aqueous infitration.

14. Theories of volcanic action—seat of volcanic fires.

15. Volcanic products, unconnected with any
existing crater—their origin determined :
(1.) By their mineral structure ;
(2.) By their geological relations ;
(3.) From the effects produced by their
junction with other rocks.
16. Application of these rules to formations of
Basalt—Examples—Basaltic Rocks of Auvergne,
&c.
17. Effects of volcanic action—changes in the
relative level of land and water. Examples.
18. Proofs of igneous action among rocks of
all epochs in the history of the earth.

CHAP. IV.

*Ancient alluvion ("Diluvial detritus")—including
all superficial transported aqueous deposits
which are unconnected with the present
mechanical action of the waters.*

1. This alluvion distinguished generally from
alluvial detritus by its antiquity—position—organic
remains—not being in progress of formation,
&c. &c.
2. Proofs of great changes in the level of
coasts during various geological periods—of the
elevation of mountain chains at successive periods.

3. Effects of these movements upon the waters.

4. Proofs of the former action of great denuding currents—forms of mountains—excavation of valleys—position of gorges, &c.—singular relations of many river channels to the elevated regions of the country.

5. Further proofs from the nature and position of the transported materials. Examples—

(1.) Boulders of the Alps;

(2.) Of the north-eastern plains of Europe;

(3.) Of the north of England.

6. Stratified masses of old alluvion—successive epochs of its formation on the flanks of the Alps, &c.

7. *Organic remains.* — (1.) In rolled masses derived from older formations.—(2.) Land and marine shells.—(3.) Bones of *mammalia* of extinct and living species, &c. &c.—No human bones.(?)

8. Description of some remarkable species.

9. Great local deposits of bones, formed before or during the period of the ancient detritus. Examples.

10. Ossiferous Caverns, Osseous Brucias, &c.

PART II.

CHAP. 1.

§ 1. *Internal structure of the earth.*

1. Observations by which the internal structure of the earth is discovered.

2. *Strata*—Their relations to each other—conformable — unconformable — strike or direction — dip—inclination. Examples.

3. *Formations*—natural groups having a common origin during one uninterrupted period.

4. Materials of which strata are formed — chemical—mechanical—or mixed.

5. Elements which enter into the composition of rocks—great variety of natural combinations— Geological descriptions confined to such as are elaborated on a great scale.

6. *Simple minerals* which enter into the composition of crystalline rocks—modes of determining their species.

7. Examples — Quartz — Felspar—Mica—Talc —Chlorite—Hornblende—Augite—Schorl—Calc-spar, &c. &c.

C

8. *Rocks*—simple and compound.
9. *Structure*— granular—slaty—porphyritic—amygdaloidal—columnar—globular, &c.
10. Rocks of double structure.
11. Cemented rocks—of regular and irregular structure.
12. *Stratified and unstratified.*—Conclusion.

§ 2. *Organic remains—state in which they occur in various orders of strata.*

1. Ancient theories connected with their appearance.
2. *Petrifaction*—conditions necessary to it.
3. Some of the principal classes of organic remains.
   (1.) *Plantæ*—different modes of their preservation.
   (2.) *Zoophyta* — sponges — alcyonia — many genera of hard stony corals, &c.
   (3.) *Radiaria* — Encrinites—Pentacrinites—Echinites, &c.
   (4.) *Testacea*—Bivalves—Univalves—single-chambered and multilocular.
   (5.) *Crustacea.*
   (6.) *Pisces.*
   (7.) *Reptilia.*
   (8.) *Aves.*
   (9.) *Mammalia.*

4. Importance of organic remains — in the identification of contemporaneous deposits — in determining the successive conditions of the earth.

Chap. II.

Classification of stratified rocks.

1. *Stratified rocks* — Primary — secondary — tertiary.

(*a.*) *Tertiary class* — including all regular deposits superior to the Chalk, and older than the old alluvion of the neighbouring regions.

(*b.*) *Secondary class* — all regular deposits, commencing with the chalk and ending with the old red sandstone.

(*c.*) *Primary class* — all stratified rocks not included in the two preceding classes — subdivided into two systems of formations.

(1.) *Upper series* —— including the principal transition rocks of Werner.——In England containing subordinate beds of limestone with many organic remains.—In Ireland containing subordinate formations of coal.

(2.) *Lower series*—highly crystalline
stratified masses — generally of
slaty texture—very rarely containing traces of organic remains.
2. Origin of the term *Transition*—with what
limitations it is applicable to the upper series—
no clear line of separation between the two.

CHAP. III.

TERTIARY CLASS.

§ 1. *Paris Basin.*

1. Origin of the term, *tertiary,* and present extension of it.
2. *Paris basin*—natural separation of the deposits into three principal groups.
3. Order of superposition in the Paris basin—
A series of horizontal deposits resting on an
irregular water-worn surface of chalk.

Ascending order as follows :
(1.) Plastic clay and sand—occasionally containing lignite—Interspersed fresh-water
shells, &c.
(2.) *Calcaire grossier.* — Chloritic sand—
coarse shell limestone—marl—clay, &c.
—Innumerable marine shells — rarely
freshwater shells—very rarely bones of
mammalia.

(*3.*) Siliceous limestone—sometimes passing into a variety of buhr-stone—partially covering and sometimes replacing the preceding group—rarely with freshwater shells.

(*4.*) Gypseous deposits—subdivided into three systems of beds.

   (*a.*) Bands of indurated marl—slaty marl with menilite and gypsum —the lowest portion containing some marine shells.

   (*b.*) Gypsum beds with a little marl —Fossil fish—no fossil shells.

   (*c.*) Great Gypsum beds—lower part impure & siliceous—middle prismatic — upper part alternating with marls.—Innumerable bones of mammalia, frequently of extinct genera — bones of birds, fishes, and turtles.

(5.) Marl—separated into two formations :

   (*a.*) White calcareous freshwater marl —resembling that in the next inferior group.—Silicified remains of palm trees—freshwater shells —&c.

   (*b.*) Marine marls—yellow, with fish bones and marine shells—green marls with balls of sulphate of strontian ——yellow argillaceous

marls with fish-bones and marine shells——The whole system
surmounted by two oyster beds
separated by a bed of white marl
without shells.

(6.) Sand and Grit of great thickness—Siliceous beds, nodules, and calcareous concretions. In the upper portion concretions of various colours, sometimes calcareous—casts of marine shells, &c.

(7.) Buhr-stone—freshwater limestone, &c.

(*a.*) Green or red sand and marl with
beds and concretions of buhr-
stone *(meulière)*—without shells.

(*b.*) Beds of compact or earthy freshwater limestone—siliceous beds
and concretions — resinous flint
—buhr-stone, &c.—Gyrogonites,
many freshwater univalves, &c.

4. General conclusions—separation of the preceding groups into marine and freshwater
formations.

5. Other freshwater deposits of central France
—extraordinary association with volcanic rocks,
&c.

§ 2.

Basins of London and the Isle of Wight.

1. These basins once probably continuous —
Great dislocations of the chalk in the southern
portions of England—Outlying tertiary masses.
2. Successive deposits in the ascending order.
(1.) A great complex marine deposit subdivided into—

(*a.*) *Sand and plastic clay*—Sands of
various colours, alternating with
bands of grit—beds of pebbles
—clay with calcareous concretions, *septaria,* and marine shells
—Subordinate bands of potter's
clay, lignite, &c. Sometimes
beds of marine and freshwater
shells, mixed or comminated near
the bottom of the series.

(*b.*) *London clay*—Clay with *septaria*
—calcareous sand-stone & green
earth —— rarely beds of coarse
shell-limestone —&c. Numerous
remains of marine shells—fish—
turtles—crustacea—wood—seed
vessels, &c. &c.

(*c.*) Sand with siliceous concretions,
&c. Bagshot sand.

(2.) Great complex deposit of shell limestone, indurated marl, unctious marl, clay, sand, &c.—generally lacustrine, but in the Isle of Wight containing subordinate argillaceous beds with a mixture of marine and freshwater shells.—*Gyrogonites* and other remains of plants—remains of fish—rarely bones of *mammalia*.

3. Comparison of the deposits of the basins of Paris and the Isle of Wight.

§ 3.
Upper tertiary groups.

1. *Crag deposits* on the coasts of Suffolk and Norfolk—corresponding deposits on the Loire—Proofs that these deposits are of a newer order than any of the preceding groups, derived from sections and groups of fossils.—Remains of fish—rolled bones of *mammalia*—shells of living and extinct species—corals, &c. &c.

2. Tertiary groups of the Sub-Apennine regions—of the Eastern Alps, &c. &c.

3. Independence of different tertiary deposits whether marine or freshwater—difficulties in their classification—principles on which it is attempted—Examples.

4. Deposits of numerous marine shells of existing species at various elevations—form a connecting link between tertiary and modern deposits.

5. Connexion of some of the newer tertiary deposits with rocks of volcanic origin. Examples.

Chap. IV.

Secondary Class—subdivided into many groups of formations.

§ 1.

Cretaceous series—subdivided into the complex groups of the Chalk and Green sand.

1. *Chalk.*
   (1.) In England this group is subdivided into—
   - (*a.*) Upper chalk — with numerous flints.
   - (*b.*) Lower chalk—with interspersed flints.
   - (*c.*) Chalk marl—grey chalk—without flints.

   Examples of these subdivisions in various natural sections.
   (2.) Varieties of structure, colour, &c.— magnesian chalk — indurated chalk— granular chalk in contact with trap.
   (3.) Range — elevation — inclination—thickness —remarkable localities—soils— springs, &c.
   (4.) Organic remains.

D

2. *Green-sand group.*

    (1.) In England this group is subdivided into—

        (*a.*) Upper green-sand—malm-rock—fire-stone.

        (*b.*) Galt—of Cambridge & Folkstone.

        (*c.*) Lower green sand—Woburn sand —Kentish rag, &c.

    (2.) Varieties of mineral structure—changes from the suppression of one or more of the subdivisions—passage into and alternation with chalk marl—In the lowest part of the third division (*c.*) shelly beds resembling portions of the oolitic series.

    (3.) Range—extent—remarkable localities.

    (4.) Organic remains.

3. Cretaceous series of the Continent—*Pläner Kalk. Quader Sandstein.*

4. General Geographical distribution of this series—its character and elevation in the chains of the Alps and Pyrenees. Changes of structure, inclination, &c.

5. On certain deposits which have a character intermediate between secondary & tertiary groups. Examples — Maestricht—Shores of the Baltic— Valley of Gosau in the Eastern Alps, &c.

6. General character of the organic remains in the cretaceous series.—Conclusion.

§ 2.

Weald-clay, Hastings-sand, and Purbeck lime-stone—A great complex deposit, chiefly of fresh-water origin, breaking the continuity of the British secondary series.

1. *Weald Clay, (Oak-tree Clay.)*

    (1.) Varieties of structure—subordinate bands of shell limestone (Petworth marble) with numerous *Paludinæ*—bands of grit covered with the *Cypris faba*——iron-stone, &c. — in the lower part separated by numerous bands of grit more or less ferruginous and passing into the lower group.

    (2.) Fossils, chiefly freshwater—alternations of marine and freshwater shells near the upper surface.

2. *Hastings-Sand, or Iron-Sand.*

    (1.) Varieties of structure

        (*a.*) Grey calciferous sandstone some-times passing into a conglome-rate—sandy clay, &c.

        (*b.*) Sandstone—often friable and of a yellowish-white colour—some-times ferruginous.

        c.) Sandstone with calcareous con-cretions—ferruginous sandstone

—layers of iron stone—variegated sandy clay—shale—Fragments of carbonized wood —— lignite, &c.

(*d.*) Argillaceous shelly limestone alternating with slaty argillaceous marls—abounding in shells of the genus Cyclas, &c. Forms a passage into the next group (?)

(2.) Fossils. — Plants — shells chiefly freshwater—Fish of freshwater genera —— Crustacea — Gigantic carniverous and herbiverous reptiles—Pterodactyls—&c.

3. *Purbeck limestone.*

(1.) Mineral structure—Marble with numerous Paludinæ—shelly calcareous flagstone—indurated marl—layers of argillaceous marl—shale—&c.

(2.) Fossils.—Numerous impressions of fish — fragments of turtles. — Freshwater and marine shells.

4. Limited extent of the three preceding groups —Contemporaneous deposits on the Continent.

5. Condition of the region now occupied by the south coast of England between the period of the oolites and the commencement of the preceding series.——*Dirt bed* of the Isle of Portland —— *Cycadeoideæ—Coniferæ* in an erect position.— Conclusion.

§ 3.

Oolitic series.

1. Place of this series defined by natural sections—character of the oolitic terraces—continuity of the great oolitic terrace.

2. The series in the south of England made up of three groups of deposits containing oolitic limestone, each resting on a great bed of clay.

3. *Upper group—Portland Oolite, and Kimmeridge Clay :*

   (1.) *Portland oolite formation.*

       (*a.*) Thin beds of limestone—compact, earthy, and cellular—chert, &c. Very few fossils.

       (*b.*) Strong beds of oolitic limestone. —mixed and alternating with masses containing numerous casts of shells—*Turritellæ, Trigoniæ,* &c. &c.

       (*c.*) Beds of varied structure—hard, brown, splintery, cherty, &c. —White earthy beds with black flinty concretions—Fossils.

   (2.) *Kimmeridge clay.*

       (*a.*) Sand and gritty beds alternating with clay (Kimmeridge sand).

    (*b.*) Clay——shale, sometimes highly bituminous— numerous fossils— *Ostrea deltoidea*, *Gryphæa virgula*—&c. &c.

    (*c.*) Clay—sometimes mixed with green earth—alternating with the upper calc-grit of the next inferior group—*Ostrea deltoidea*, &c. &c.

  (3.) Continuity of the Kimmeridge clay—gradual thinning of the Portland rock in its range towards the north.

  (4.) Varieties of mineral structure—range—inclination—elevation—thickness, &c.

4. *Middle group—Coral Rag Oolite and Oxford Clay.*

  (1.) *Coral Rag Oolite.*

    (*a.*) Irregular beds of shelly calc-grit—often ferruginous — sometimes alternating with the lower beds of Kimmeridge clay.

    (*b.*) Broken semicrystalline and earthy beds of limestone, sometimes alternating or mixed with oolitic marl — abounding in shells and corals (Coral rag.)

    (*c.*) Oolitic freestone — often coarse and earthy—fragments of shells, &c.—Mixed or alternating with masses like the preceding subdivision (*b.*)

(*d.*) Beds of calcareous-grit—sand and sandstone, sometimes highly ferruginous, &c.—Changes in the several sub-divisions — gradual thinning of the formation in its range towards the north —- their disappearance in a part of Cambridgeshire and Lincolnshire — re-appearance in Yorkshire.—Remarkable fossils.

(2.) *Oxford clay.*

    (*a.*) Great clay, with the *Gryphæa dilatata*—bands of sandstone and calc-grit—bituminous beds—septaria, &c.

    (*b.*) Irregular beds of micaceous sandstone, calc-grit, &c.—(Kelloway rock.)

(3.) Continuity of this formation—thickness —remarkable localities—modification of its structure in Yorkshire, &c.

5. *Lower or great oolitic group and Lias Clay.*

(1.) *Great Oolitic group.*

    (*a.*) Cornbrash—earthy, loose, amorphous beds of yellow, grey, and bluish shelly limestone.

    (*b.*) Slaty calcareous sandstone alternating with sand.

    (*c.*) Yeovil marble (Forest marble)— coarse slaty shelly limestone,

sometimes alternating with beds
of clay—marl—slaty calc-grit,&c.

(*d.*) Bradford clay.

(*e.*) Great Oolite—Oolitic freestone of
Bath, Ketton, &c.

(*f.*) Fuller's earth and sand.

(*g.*) Inferior Oolite—Coarse brown
oolitic limestone—calcareous concretions—sand, &c.—Continuity
of this formation — thickness —
changes and modifications of
structure, by the suppression or
expansion of one or more of the
sub-divisions. Examples.

(2.) *Lias.*

(*a.*) Upper Lias clay.

(*b.*) Marlstone ——Gritty micaceous
marl, more or less indurated, of
a greenish or yellowish brown
colour.

(*c.*) Lower Lias clay.

(*d.*) Blue and white Lias—with *Gryphæa incurva,* &c.—Continuity
of the formation — thickness —
changes of structure—first appearance and gradual expansion
of division (*a.*)—Modification of
division (*b.*)—Minerals—mineral
springs — extraordinary fossils,
&c.

6. *Oolitic series of the Yorkshire Coast.*

(1.) **Kimmeridge clay.**

(2.) **Coral rag oolite, calc-grit, &c.**

(3.) **Oxford clay and Kelloway rock.**

(4.) **A complex carboniferous series, in which the types of the lower oolitic group of the south of England are almost entirely lost**—subdivided into :

(*a*) Sandstone, shale, iron-stone, carbonaceous beds,&c.—surmounted by a band of impure shell limestone, sometimes oolitic.

(*b*.) Impure limestone, sometimes oolitic—supposed from its fossils to be on the parallel of the Bath oolite.

(*c*.) Sandstone, shale, coal, &c.— resting on bands of shelly calcareo-ferruginous sandstone.

(5.) Lias—subdivided into :

(*a*.) Upper lias shale (alum shale.)

(*b*.) Shale, iron-stone, and micaceous sandstone (marl-stone.)

(*c*.) Lower lias shale and lias — *Gryphæa incurva*, &c.

7. Re-appearance of the carboniferous groups of the oolitic series in Sutherland and the Hebrides —extraordinary associations with basaltic rocks— coast-sections of the Isles of Sky and Mull.

8. Great development of the oolitic series in France.

E

(1.) Section on the coast of Normandy.

    (*a.*) *Argile de Honfleur* (Kimmeridge clay.)

    (*b.*) Hard cherty limestone — coralline oolite—calc-grit, &c. (Coral rag or middle oolite group.)

    (*c.*) *Argile de Dives* (Oxford clay.)

    (*d.*) Oolitic fissile shell limestone — Caen freestone — yellow calcareous sandstone——ferruginous oolite, &c. (Group of the great oolite.)

    (*e.*) Lias—*Gryphæa incurva*, &c.

(2.) Section of the Oolites of Burgundy and the north-eastern Provinces.

    (*a.*) Light coloured compact limestone. (Portland rock.)

    (*b.*) Grey marls and marly limestone, with the *Gryphæa virgula*. (Kimmeridge clay.)

    (*c.*) Compact and earthy limestone—coralline oolite, &c. (Coral rag.)

    (*d.*) Grey marly limestone and calc-grit, with the *Gryphæa dilatata*. (Oxford clay.)

    (*e.*) Great plateau of oolitic limestone—yellowish marly limestone — entrochite limestone, &c. — (Group of the great oolite.)

    (*f.*) Marl and Gryphite limestone. (Lias.)

§ 4.

New red sandstone series.

2. Relations of the series to the inferior and superior deposits exhibited in various natural sections.

3. Subdivisions of the whole series.

(1.) *Yorkshire and Durham.*

(*a*) Upper red, or variegated marl and gypsum.

(*b*.) Red and variegated sandstone—-with subordiate masses of red marl and gypsum.

(*c*.) Thin bedded grey limestone— occasionally yellow cellular and magnesian—rarely with organic remains.

(*d*.) Lower red marl and gypsum.

(*e*.) Magnesian limestone—crystalline compact —— earthy —— cellular— oolitic—fetid—&c.——Occasionally with organic remains — *Producta, Spirifer, Gorgonia, Retipora, &c.*

(*f*.) Thin bedded compact limestone (with *Producta, Spirifer,* &c.), resting on marl slate with impressions of plants, and fish (*Palæothrissum,* &c.)

(*g*.) Lower red sandstone—generally unconformable to the carboniferous series— sometimes conformable and appearing to pass

into the upper coal-grits — rarely
with remains of plants.

(2.) *Coast of Cumberland.*

(*a.*) Red and variegated sandstone
resting on red gypseous marls.

(*b.*) Magnesian limestone — cellular,
earthy, fetid, &c.

(*c.*) Magnesian conglomerate — containing rolled masses of mountain limestone.

(*d.*) Lower red sandstone—of considerable thickness — unconformable to the coal measures—a few
stems of plants.

(3) *Succession in a part of Shropshire*—

(*a.*) Red or variegated marl and
gypsum.

(*b.*) Red and variegated sandstone.

(*c.*) Magnesian conglomorate — with
rolled masses of mountain limestone.

(*d.*) Lower red sandstone—of considerable thickness, and apparently parallel to the beds of the coal
measures on which it rests.

(4.) *Bristol Coal Fields and Valley of the Ex.*

(*a.*) Red or variegated marl, gypsum,
&c.

(*b.*) Red and variegated sandstone.

(*c.*) Magnesian conglomerate with
rolled masses of mountain limestone—no traces of the lower
red sandstone as a distinct deposit
—the whole series resting hori-
zontally on the inclined beds of
the older formations.

4. External character of the red sandstone
country — changes of structure — forest-sand —
conglomerates —thickness — minerals—cobalt,
manganese, calamine, rock salt, &c.

5. External character of the magnesian limestone—varieties— remarkable localities—minerals
and mineral springs—soils—&c.

6. Great expansion of this series in Europe—
constancy of minerals and fossils—exceptions.

7. Series in Central Germany and Eastern
France with their British equivalents.

(1.) *Keuper* or *Marnes irisées* — Highest(greenish red) marls of the upper red marl & gypsum?

(2.) *Muschel Kalk* — fossil limestone of Lunéville. — Not discovered in the British series.

(3.) *Bunter-sandstein* —*Grès bigarré* or *Grès rouge*. — Red & variegated sandstone — new red sandstone.

(4.) *Rauchwacké, Zech-stein,* &c. — not much developed in Eastern France. — Magnesian lime-stone group.

(5.) *Kupfer Schiefer.* { Marl-slate of Durham, with fish, &c.

(6.) *Rothe-todte-liegende — Grès des Vosges.* { Lower red sandstone group.

8. Rock salt—in the Alps, Poland, &c.—not confined to the new red sandstone series—theories of its formation—Conclusion.

§ 5.

Great carboniferous series, including the mountain limestone.

1. General character of the series—alternations of limestone, sandstone, shale, ironstone, &c. with subordinate bands of coal.

2. Relations of this series to the inferior and superior groups exhibited in sections.

3. Varieties of limestone—crystalline—compact —earthy—magnesian—bituminous—fetid—rarely oolitic—beds abounding in encrinites, shells, and corals, &c.

4. Varieties of sandstone—millstone grit—freestone—micaceous slaty sandstone—chert—beds more or less ferruginous, often marked and penetrated by leaves and stems of plants.

5. Varieties of shale—common—bituminous, sometimes passing into coal—sandy and micaceous,

passing into bands of sandstone — calcareous
and indurated *(calp)*—passing into shale-limestone,
rottenstone, crow-limestone, &c.——subordinate
beds and bands of ironstone—numerous impressions of plants.

6. Varieties of coal—bituminous shale—bituminous coal — anthracite——mineral charcoal——
lignite—&c.—Impressions of carbonized plants.

7. Great extent of this series in the British Isles
—mode in which the several members are grouped
together in different coal fields—Examples.

*Coal fields of Somersetshire, Gloucestershire,
and South Wales.*

(1.) Upper and productive part of the series,
subdivided into :—

(*a.*) Upper coal field ;

(*b.*) Pennant grit ;

(*c.*) Lower coal field ;

Separated generally from the lower un-
productive calcareous groups by beds of
coarse sandstone—mill-stone grit—&c.

(2.) Great mountain limestone group— un-
productive—subordinate beds of sandstone and shale—in the lower part generally separated from the old red sandstone by a deposit of shale—This group
generally arranged in broken or continuous ridges—highly inclined—dipping
towards a common centre, and forming
the base of distinct coal-basins.

Carboniferous series of Derbyshire.

(1.) Great productive coal-field, surmounted
by the terrace of the magnesian limestone.

(2.) Mill-stone grit.

(3.) Great shale and shale-limestone.

(4.) Mountain limestone, interrupted by three
great irregular beds or tabular masses of
trap—the base of the series not exposed.

*Carboniferous series of Yorkshire and West-
morland.*

(1.) Great upper coal field.

(2.) Complex group of millstone grit shale
and coal, in several alternations—sometimes with bands of impure crow-limestone.

(3.) Complex group, in which five deposits
of limestone alternate with sandstone
containing impressions of plants, and
with beds of shale, more or less bituminous, containing two or three productive beds of coal, &c.

(4.) Great scar-limestone with a few subordinate bands of shale and sandstone—
separated sometimes from the inferior
system by a deposit of bituminous shale.

The whole series sometimes rests immediately upon the inclined edges of
greywacké—or conformably upon great
beds of old red conglomerate.

F

Carboniferous series of Northumberland.
(1.) Great productive upper coal field.
(2.) Great complex group—including mill-
stone grit—limestone—shale with bands
of impure coal — tabular masses and
dykes of trap, &c.
(3.) Lower productive coal-field with subordinate beds of limestone forming the
base of the whole series—resting unconformably on greywacké—separated
from the porphyry of the Cheviots by
old red conglomerate.

Carboniferous series of Ireland.
(1.) Great complex contorted groups on
the parallel of the upper English coal
field.
(2.) Widely extended horizontal deposits of
mountain limestone resting conformably
on old red sandstone.

Carboniferous series of Scotland.
(1.) Coal formations between the Firths of
Forth and Clyde, on the parallel of
the lower groups of Northumberland?
(2.) Series of the Isle of Arran, &c.
8. Remarkable foreign localities—absence of
the formation in the Alps and Pyrenees—appearance of the coal fossils on the parallel of the lias.
9. Origin of coal—comparison with some of

the newer deposits of *lignite*—enumeration of the most remarkable coal plants.

10. Enormous extent and rate of the present excavations in the British Isles—first commencement—probable duration, &c.

11. External character of coal countries — minerals—mineral springs.

12. External character of mountain-limestone countries—minerals and mineral waters—ravines —caverns—intermitting springs—soils—&c.

13. Characteristic fossils—Conclusion.

§ 6.

Old Red Sandstone and Conglomerate.

1. Great extent of this group in the British Isles—Its relations to the inferior and superior systems exhibited in various natural sections.

2. Mineral structure —— subordinate beds of concretionary limestone *(cornstone)*.

3. In the west of England appears to graduate into greywacké—In the north of England, chiefly composed of conglomerates resting unconformably on the older strata.

4. Old conglomerates of the coast of Sutherland—conglomerate, bituminous schist, and red
sandstone of Caithness — fossil fish — crustacea,
&c.

5. Great extent of the formation in Ireland.

6. Obscure development of this group in many
parts of the Continent.   Conclusion.

Chap. V.

Primary stratified rocks.

§ 1.   Introduction.

1. Prevailing characters of primary rocks—slaty
—highly inclined—contorted—more or less crystalline.   Examples.

    (1.) Clay slate, greywacké slate, greywacké,
        quartz rock ;
    (2.) Chlorite slate ;
    (3.) Talc-slate ;
    (4.) Mica slate ;
    (5.) Hornblende slate ;
    (6.) Gneiss,
        &c. &c.

2. Separation of the primary rocks into an upper and lower series, each composed of many groups.

3. The groups complex and ill defined—the order of succession sometimes indefinite—examples derived from actual sections.

4. Beds of the upper groups generally unconformable to the secondary systems—sometimes conformable to, and appearing to graduate into, the old red sandstone. Examples.

5. Junction of the various groups with crystalline unstratified rocks forming the axes of primary chains—great complexity of structure—penetration of granitoid veins, &c. Conclusions.

6. General relations of stratified primary rocks exhibited in transverse sections of remarkable primary chains.

§ 2.

Upper series of primary stratified rocks, subdivided into three groups.

1. *First group*—Greywacké slate, Greywacké, Clay-slate, &c.—with subordinate beds of limestone, containing numerous organic remains.

(1.) Varieties of structure.

Rocks of slaty texture.

(*a.*) Coarse greywacké slate or flagstone—exfoliating parallel to the planes of deposit.

    (*b.*) Fine greywacké slate and clay
        slate—cleaving into fine laminæ
        which range parallel to the *strike*
        of the beds, but are inclined to
        them at various angles.

    (*c.*) Other varieties depending upon
        the different degrees of indura-
        tion—alum slate, whetstone slate
        —flinty slate, &c.

Rocks not of slaty texture.

    (*a.*) Conglomerates passing into mill-
        stone and coarse sandstone ;

    (*b.*) Greywacké ;

    (*c.*) Quartz rock—examples (in the
        Hebrides and North-western
        Highlands) of this rock alter-
        nating with greywacké slate, and
        forming a passage into the old
        red sandstone.

(2.) General diffusion of calcareous matter
    through this group—beds highly calca-
    reous passing into pure limestone —
    ranging parallel to the strike of the group.

(3.) Prevailing characters of this limestone
    in Cornwall, Devonshire, Somersetshire,
    Gloucestershire, Wales, Cumberland, &c.

(4.) Organic remains—rarely diffused through
    the ordinary beds of the group in the
    form of casts—abundant and well pre-

served in the calcareous beds. Examples. Corals of the genera, Caryophyllia, Astrea, Catenipora, &c.—Crinoeidea — Shells of the genera, Orthoceras, Producta, Spirifer, Terebratula, &c.—Trilobites—Bones of fish.

(5.) Vegetable remains.—anthracite—in Ireland beds of pyritous shale with subordinate beds of coal — impressions of Equiseta and Calamites.

(6.) Range and extent of this group in the British Isles—dislocations — contortions —— elevation —— minerals——porphyritic dykes—remarkable localities British and foreign—&c.

2. *Central group*—Clay slate—greywacké slate, &c.—alternating with felspathic and porphyritic rocks.

(1.) Great development of this group in the British Isles. Examples.

(a.) *Central schistose group of Cumberland, Westmorland, and Lancashire,* composed as follows, viz.: Various porphyritic rocks with a base of compact felspar; hornstone, hornstone slate; greenstone; concretionary beds and pseudo-conglomerates; coarse breceiated masses passing into and blended with porphyries, &c. &c.—alternating indefinitely with coarse quartzose chloritic beds, more or less fissile ; and with fine quartzose and chloritic masses cleaving into thin laminæ parallel to the strike, but inclined to the planes of dip, &c.—the whole system stratified, having a remarkable constancy in its strike and dip, and without traces of organic remains—In Cumberland this group sometimes rests immediately upon the *third schistose group*—sometimes upon granite or syenite, irregularly protruded between the second and third groups.

(*b.*) *Central schistose group of North Wales.* In Caernarvonshire and Merionethshire composed as follows, viz. :—Various porphyritic rocks with a base of compact felspar ; greenstone in beds and irregular masses ; felspathic and quartzose beds in globular concretions ; pseudo-conglomerates, and irregularly breceiated masses passing into porphyry, &c. &c.—alternating indefinitely, with quartzose and chloritic slaty masses ; quartzose beds passing into granular quartz rock ; felspathic slate, breceiated or in minute contortions; talc slate passing into porphyry ; greywacké with casts of Terebratulæ, very rarely with impressions of corals ; great masses of greywacké slate and clay slate, both soft and indurated,

and with a cleavage inclined to the
planes of dip, &c.—the whole series
based upon greywacké-slate and clay-
slate alternating with two or three
zones of porphyritic and felspathic rocks.
—In Caernarvonshire this system is bent
into great curves producing a series of
anticlinal lines parallel to the bearing
of the chain; and in some places abuts
against, or is irregularly interrupted
by, protruding masses of syenite and
porphyry.

(2.) Other localities—obscure development
of this group in many primary chains—
its upper and lower limits ill defined.

(3.) Elevation & external character—porphyritic dykes—minerals—mineral springs.

3. *Third group.*—Fine glossy clay-slate—greywacké-slate—greywacké—&c.

(1.) Examples of this group.

(a.) *Cumberland*—Soft dark-coloured glossy
clay-slate (Skiddaw-slate) — greywacké-
slate with subordinate beds of coarse
greywacké passing into sandstone, &c.
—This group contains numerous quartz
veins — does not generally effervesce
with acids—is indurated and altered in
structure near its junction with mountain masses of porphyry and syenite,

G

by which it is penetrated, traversed and overlaid—contains no organic remains—rests on chiastolite slate, through which it passes into an inferior crystalline group.

(b.) *North Wales*—An obscure group, composed of greywacké-slate and greywacké associated with porphyry, extending from the western base of Snowdonia to the Menai straits (?)

(c.) *Cornwall*—the central and lower portions of the *killas*—associated with porphyry dykes.

(2.) Other localities—varieties of structure—minerals—mineral springs.

4. Great extension of the preceding series—external character of the regions it traverses—extensive changes of mineral structure and ill-defined limits of the several groups in the vicinity of unstratified rocks.—Examples.

§ 2.

Lower series of primary stratified rocks.

1. Crystalline and slaty structure of the prevailing groups in this series.

2. Examples of highly-crystalline rocks (resembling those of the lower primary groups,) associated with the upper primary series, and with secondary deposits—generally in the vicinity of unstratified rocks.

(1.) Hornblende-slate and mica-slate of the Lizard (associated with serpentine,greenstone, diallage-rock, &c.) overlying *killas* and greywacké.

(2.) Felspathic slate, greenstone and hornblende-slate, &c. — in various parts of Cornwall near the junction of the granite and *killas*.

(3.) White crystalline marble of the Isle of Sky, at the junction of Sienite mountains with lias.

(4.) Highly crystalline rocks among the secondary and newer primary groups of the Alps.

Hence the difficulty of determining the epoch of the older groups by their mineral structure.

3. Only two examples of the lower primary
groups in England and Wales.

(1.) A group interposed between the clay
slate of Skiddaw and the central granite
of Skiddaw forest—composed nearly as
follows :

(*a.*) Clay slate with chiastolite.

(*b.*) Hornblende-slate mixed with slate
made up of matted crystals of
chiastolite.

(*c.*) Mica-slate—-finely. laminated—
quartzose and coarsely laminated
—&c.

(*d.*) Irregularly striped or laminated
rocks approaching the character
of gneiss.

(2.) A group extensively developed in the
Isle of Anglesea and on the south-west
coast of Caernarvonshire, composed prin-
cipally of chlorite slate.

(*a.*) Chlorite-slate — finely laminated
—quartzose—contorted —some-
times micaceous, and passing into
quartzose mica-slate—&c.

(*b.*) Contorted quartz-rock—separated
into slaty masses by flakes of mica
ranged in parallel planes inclined
to the planes of stratification
(Holyhead Mount).

(*c.*) Chlorite slate penetrated by veins of carbonate of lime—passing into amorphous masses of serpentine and vert antique (Anglesea) —similar masses associated with jasper (coast of Caernarvonshire).

(*d.*) Chlorite slate penetrated by veins of carbonate of lime — passing into great beds of crystalline limestone penetrated by veins of chlorite—associated with masses of jasper and serpentine (S. W. coast of Caernarvonshire).

This group associated with and traversed by many masses and dykes of trap.

4. *Chlorite-slate—talc-slate—mica-slate—horn-blende-slate.*

(1.) Great groups of strata formed out of the several species—alternating indefinitely with each other. Examples.

(2.) Subordinate rocks— serpentine —marble —dolomite—&c.

(3.) Imbedded minerals.

(4.) Contortions—elevation — decomposition —extent—remarkable localities.

5. *Quartz-rock.*

(1.) Mineral structure—remarkable varieties —great development in the Highlands of Scotland.

(2.) This rock forms a passage from the primary to the secondary system—alternates with the whole primary series—proves the existence of mechanical deposits in all parts of the series—Examples.

6. *Limestone*—common and dolomitic—generally subordinate to the other groups—sometimes alternating with them in large independent masses. —Examples.

7. *Gneiss*—alternates with all the preceding groups—sometimes subordinate—forms the prevailing rock in many extensive regions.

(1.) Mineral structure—laminar—schistose— granitic.

(2.) Varieties of structure—from the suppression or replacement of one constituent—from the entire separation of the several constituents producing beds of simple mineral structure—&c.

(3.) Subordinate mineral masses — subordinate minerals.

(4.) Subordinate veins, especially near the junction of gneiss with granite.

(5.) Elevation—contortions—range — extent —remarkable localities.

8. Remarkable successions of the primary groups in different natural sections—Conclusion.

Chap. VI.

Unstratified Rocks.

§ 1. *Introduction.*

1. Prevailing characters of rocks of this class—
modes of their distribution—

(1.) In irregular overlying masses :

(2.) In protuberances and in great tabular
masses interrupting the regular succession of the stratified rocks :

(3.) In veins :

(4.) In great central masses producing anticlinal lines and often forming the axes
of mountain chains :

(5.) In regions unconnected with stratified
rocks.

2. Great variety of minerals and mineral-struc-
ture in the rocks of this class—Examples of some
of the modifications :

(1.) Rocks of simple or nearly simple structure.

(*a.*) Claystone in various states of
induration ;

(*b.*) Clinkstone ;

(*c.*) Compact felspar ;

(*d.*) Basalt ;
(*e.*) Pitchstone ;
(*f.*) Serpentine ;
&c. &c.
(2.) Further subdivisions arising out of the
modifications of the preceding species—
Examples :
    (*a.*) Felspar porphyry ;
    (*b.*) Pitchstone porphyry ;
    (*c.*) Pearlstone ;
    (*d.*) Amygdaloids ;
    (*e.*) Trachyte ;
    (*f.*) Trachytic porphyry ;
    &c. &c.
(3.) Binary compounds, more or less distinctly
granular—Examples:
    (*a.*) Greenstone—becoming compact
      and passing into a basaltic form.
    (*b.*) Augite rock(*Dolerite,Melaphyre*)
      —often ·passing into a basaltic
      form.
    (*c.*) Diallage rock (*Euphotide*)—pas-
      sing into serpentine.
    &c. &c.
(4.) Granitoid rocks — generally composed
of more than two constituents—contain
felspar often distinctly crystallized—
Examples—granite—syenite—&c.
(5.) Other modifications—rocks of globular
and columnar structure.

3. The unstratified rocks associated with stratified rocks of different epochs—pass into and replace each other—their place in the series not defined by their structure.

4. General agreement of the series of unstratified rocks with true volcanic products.—Conclusion.

§ 2.

Unstratified rocks associated with tertiary deposits.

2. Streams of basaltic lava, and masses of volcanic breccia, following the course of existing valleys—sometimes connected with existing craters. Examples—valleys of Auvergne and Viverais—valleys on the right bank of the Rhine near Andernach—&c.

2. Portions of more ancient streams of lava anterior to the existing valleys, masses of volcanic brecia, &c., overlying and sometimes alternating with tertiary strata.—Examples.

(1.) Older basaltic rocks of Auvergne and Viverais.

(2.) Volcanic breccias and basaltic lavas overlying and sometimes alternating with the newer tertiary deposits of Styria.

(3.) Basaltic plateaux of Hessia.

3. Great masses and protuberances, more or less granitoid and porphyritic—sometimes overlying

H

tertiary deposits—sometimes producing anticlinal lines.—Examples :

   (1.) Trachytic domes of Auvergne.

   (2.) Trachytic masses of Styria and Hungary.

   (3.) Trachytic masses on the right bank of the Rhine—(Drachenfels.)

4. Subaerial and subaqueous volcanic products —resulting varieties of trap rocks.

§ 3.

Trap rocks associated with secondary strata.

1. Overlying masses—dykes—irregular tabular masses alternating with secondary strata.—Examples in detailed sections.

   (1.) North of Ireland.

   (2.) Hebrides.

   (3.) Derbyshire, Durham, and Northumberland.

2. Great complications of mineral structure—Examples from the Isle of Arran.

3. Effects produced on various secondary deposits by the intrusion of trap.—Examples :

   (1.) Siliceous schist (indurated shale)—North of Ireland ; Isle of Sky ; Isle of Mull.

   (2.) Granular marble — jasper — irregularly crystallized masses with crystals of garnet and analcime, &c.—Isle of Anglesea— Teesdale.

(3.) Anthracite in contact with the trap dykes
of various coal fields.

&c. &c.

4. Mountain masses of trap—structure often
granular or porphyritic—passing into the other
varieties—Examples :

(1.) Syenite and hypersthene mountains of
the Isle of Sky—effects produced by
their contact with the lias.

(2.) Trap rocks of the valley of the Tweed—
granitic and porphyritic rocks of the
Cheviots—some of the masses prior to the
old red conglomerates—other masses of
like structure posterior to the lower portion of coal-measures—basaltic rocks.

5. Black augitic porphyry—red felspathic porphyry—serpentine, &c.—many epochs of eruption
—difficulty of defining any of them by mineral
structure.

§ 4.

*Unstratified rocks associated with the upper
primary groups.*

1 Dykes and irregular tabular masses interrupting the succession of the stratified rocks—
great variety of their structure.

2. Mountain masses irregularly protruded
amongst, or overlying, the stratified rocks—
Examples.

(1.) Basalt syenite and granite of Christiana overlying orthoceratite-limestone.

(2.) Syenite and porphyry of Cumberland irregularly traversing and overlying the Skiddaw slate.

(3.) Granite of Shap protruded between the upper and central stratified group of the Cumbrian mountains.

(4.) Syenite of the Rivals and other mountains of Caernarvonshire.

(5.) Serpentine of the Lizard — associated with porphyry, syenite. hornblende-slate, &c.—overlying greywacké slate.

3. Mountain masses producing anticlinal lines and forming the axes of primary chains. — Examples :

(1.) Granitic chain of Cornwall—relations of the crystalline central masses to the slate rocks—varieties of mineral structure—different periods of production.

(*a.*) The great masses of central granite posterior to the slate.

(*b.*) Dykes and masses of porphyry (*elvans*) posterior to the central granite.

(2.) Central granite of the Hartz—its relation to the surrounding formations—proofs of its recent protrusion.

&c. &c.

§ 5.

Crystalline masses interrupting or forming the base of the lower primary groups.

1. Great complexity of the lowest stratified groups—modes of their association with unstratified masses exhibited in sections.

2. Crystalline structure of some of the stratified groups superinduced by the action of the unstratified—examples of remarkable changes of structure probably produced by the action of granitic masses.

3. Granitic veins traversing primary strata—circumstances under which they are most abundant—theory of their origin—various examples from the primary mountains of Scotland.

4. Central granite—examples of its association with the lowest stratified groups.

5. Varieties of granite:

   (1.) From the changes of aggregation—porphyritic—globular—graphic—&c.;

   (2.) From the addition of a new constituent;

   (3.) From the suppression or replacement of a constituent. Examples.

6. Decomposition of granite—extent of granitic countries—remarkable localities.

7. Different epochs of the production of granite—not defined by the structure of the rock—granite veins—proofs of their injection in a fluid state exhibited in detailed sections.

Chap. VII.

Successive periods of elevation—metalliferous
veins—theories of the earth.

1. Review of the relations of stratified and unstratified rocks.

2. Unstratified masses generally the result of igneous action—produced during many successive periods—in different degrees of fusion—sometimes protruded in a solid form.

3. Different periods of elevation—their influence on the physical geography of Europe.

4. Geological connexions of some of the principal mountain chains of Europe—manner of investigating their several periods of elevation—sketch of the theory of M. Elie de Beaumont.

5. Physical geography of the British Isles connected with different periods of elevation.

6. Classification of some of the great faults traversing the British stratified deposits—different modes in which they are filled.

7. Smaller faults produced during periods of elevation—metalliferous veins—theories of their formation—&c.

8. Sacred theories of Burnet, Woodward, and other writers—formation of geological collections —importance of the old geological cabinets in the University of Cambridge.

9. Theories of Werner and Hutton.

10. Present condition of Geology, considered both as a practical and speculative science— desiderata—Conclusion.

FINIS.

SYLLABUS

OF

𝔄 Course

OF

BOTANICAL LECTURES.

———◄●►———

BY

REV. J. S. HENSLOW, M.A.

PROFESSOR OF BOTANY

IN THE UNIVERSITY OF CAMBRIDGE.

—··································—

Cambridge:

Printed by James Hodson, Trinity Street.

————

1828.

PREFACE.

—⸘0✣0⸘—

A few copies only of this SYLLABUS have been struck off, for the use of those gentlemen who may attend the present Course; it being the Author's intention to prepare another, containing more copious references, so soon as he shall have arranged further materials, and completed several more drawings for his Lectures.

SYLLABUS, &c.

INTRODUCTION.——Distinction between animals, vegetables, and minerals.—Organized and unorganized bodies.

Definition of a plant—method of studying botany——herbarium —vasculum—microscope —lens—chalk paper—poison, spirits of wine (1 pint) + corrosive sublimate (1 drachm.)

BOOKS.

Smith (J. E.), Introduction to Physiological and Systematical Botany, 6th ed. 8vo.
——————— English Flora, 4 vol. 8vo.
De Candolle, Thèorie élémentaire de la Botanique, 2d edit. 8vo.
——————— Organographie Végétale, 2 vol. 8vo.

PLATES.

Sowerby, English Botany, 36 vol. 8vo.
Greville, Scottish Crytogamic Flora, 5 vol. 8vo.
Curtis, Botanical Magazine. Monthly Nos.

Chemical constituents of plants.

Vegetable structure composed of,

   i. Cellular texture.

   ii. Vascular texture.

      Spiral (tracheæ), annular, dotted, beaded, and reticular vessels.

      Fibres and layers.

      Tenacity of Silk 34

      ————— Phormium tenax $23\frac{4}{5}$

      ————— Hemp .... $16\frac{1}{3}$

      ————— Flax .... $11\frac{3}{4}$

   iii. Cuticle and Epidermis.

      pores (stomata)—glands—hairs (pili) —bristles (setæ.)

   iv. Sap—proper juice—fluid and solid secretions.

Plants, cellular (cellulares) or vascular (vasculares.)

Composite organs,

   1. Conservative or fundamental,

      stem—root—leaves.

   2. Reproductive,

      flower—fruit.

N. B.—The several parts of a flower, and the method of technical description, will be explained here, to enable us to commence our "*demonstrations*" from living specimens.

Process of germination in,
1. Dicotyledons.——2. Monocotyledons.—
3. Acotyledons.

History and detailed description of each
organ.

i. Stem and branches,
collar (*collet*) between stem and root.
Herbaceous (caulis), in annual, biennial,
and perennial plants.
Woody, in under-shrubs (suffrutices),
shrubs (frutices), trees.

Composition in,
(1). DICOTYLEDONES or EXOGENÆ.
(*a*). Central system; pith—medullary
sheath—wood—alburnum.
(*b*). Cortical system; liber—outer bark
—cellular coating.
(*c*). Medullary rays.

Growth of branches, from axil of leaf.

(2). MONOCOTYLEDONES or ENDOGENÆ.
(*a*). Palms; (stipes).
(*b*). Liliaceæ, &c.
(*c*). Musaceæ.
(*d*). Gramineæ; straw (culmus).

Stems; erect, oblique, climbing, creeping,
&c.
——— solid, hollow, knotted, jointed, &c.

II. **Root.**

Compared with the stem.

Crown—caudex—rootlet (radicula)—fibrils.

Conical—fusiform—-abrupt (præmorsa)-— fibrous—tuberous—creeping, &c.

Bulbs—Tubers—Suckers.

III. **Leaves.**

Footstalk (petiolus)—limb—apex—base— margin—parenchyma—nerves: primary, secondary, &c.

Simple or compound—leaflets (foliola).

Angulinerved—penninerved—palminerved —peltinerved—pedalinerved.

Curvinerved—convergent—divergent.

Entire—lobed—toothed—notched—pinnatifid, &c.

Simply, doubly, &c. compound—pennate— palmate—peltate, &c.

Opposite—alternate—whorled, &c.

Vernation—equitant—involute—revolute— convolute, &c.

Sheath (vagina)—axil—stipule—scales— tendril—pitcher, &c.

Functions of leaves.

IV. **Flower.**

Pedicel—peduncle—bracte.

1. Calyx and sepals.

partite—divided—toothed—entire.

9

2. Corolla and petals.
   Claw—limb
      monopetalous, &c.—regular–irregular.
   Perigone (perianth)—torus—receptacle.
   Æstivation—valvular—induplicate—imbricate, &c.
3. Stamen.
      filament—anther—pollen—fovilla.
4. Pistil and carpels.
   Ovary (germen)—-ovules—-placenta——style—stigma.
5. Nectaries.
   Abortion, transformation, multiplication, and agglutination of floral organs—double flowers, &c.

Inflorescence.
   1. Axillary, indefinite, or centripetal.
   Spike—catkin (amentum)—cone (strobilus) spikelet—spadix.
   Cluster or raceme.
   Umbel—simple — compound — universal—partial.
   Head (capitulum).
   2. Terminal, or centrifugal.
   Cyme.
   3. Mixed.
   Thyrsus.
   Corymb.

v. Fruit.

    Seed—pericarp.

Carpels: epicarp, endocarp and mesocarp
—sarcocarp—-thecaphore—-placenta—
summit—-dehiscent and indehiscent—
pulp.

Follicle-—pod (legumen)-—utricle—nut—
berry (bacca)—drupe—samara—cariop-
side—achenium.

Agglutinated carpels—columella—axis—
false cells—valves—partitions—sutures.

Entire, divided, multiple, unilocular &c.
fruits.

Pouch (siliqua and silicula)—pyxidium—
capsule—pomum—cone.

Seed—horizontal—erect or ascending—inverse or descending—naked.

Skin (spermodermis): shell (testa), en-
dopleura, and mesosperm—(sarcodermis)
—umbilical chord—cicatrice (hilum)—
omphalode—micropyle—raphe—chalaze.

Kernel—amnios—albumen.

Embryo: radicle, plumule, and cotyledon
—caulicle—gemmule.

Vernation of Cotyledons—accumbent—incumbent, &c.

11

Classification of Plants.

ARTIFICIAL SYSTEM.

CLASSES.	ORDERS.
1. Monandria	Monogynia
2. Diandria	Digynia
3. Triandria	Trigynia
4. Tetrandria	Tetragynia
5. Pentandria	Pentagynia
6. Hexandria	Hexagynia
7. Heptandria	Heptagynia
8. Octandria	Octogynia
9. Enneandria	Enneagynia
10. Decandria	Decagynia
11. Dodecandria	Dodecagynia
12. Icosandria	Polygynia
13. Polyandria	

14. Didynamia   1. Gymnospermia
              2. Angiospermia

15. Tetradynamia   1. Siliculosa
                 2. Siliquosa

16. Monadelphia   Triandria
17. Diadelphia   Pentandria
18. Polyadelphia   Hexandria, &c. &c.

19. Syngenesia   1. Polygamia æqualis
                              2. ———— superflua
                              3. ———— frustranea
                              4. ———— necessaria
                              5. ———— segregata
                              6. *Monogamia*

20. Gynandria   Diandria, &c.

21. Monœcia   }   Monandria
22. Diœcia   }   Diandria, &c. &c.

23. Polygamia   1. Monœcia
                              2. Diœcia
                              3. Triœcia

24. Cryptogamia   1. Filices
                              2. Musci
                              3. *Hepaticæ*
                              4. Algæ
                              5. Fungi

NATURAL SYSTEM.

I. VASCULARES.

CLASS 1. Dicotyledones or Exogenæ.

Sub-class 1. Thalamifloræ
Orders (1 to 54.)

Sub-class 2. Calycifloræ
Orders (55 to 64.)

Sub-class 3. Corollifloræ
Orders (45, &c.)

Sub-class 4. Monoclamydeæ
Orders, &c.

CLASS 2. Monocotyledones or Endogenæ.

Sub-class 1. Phanerogamæ
Orders, &c.

Sub-class 2. Cryptogamæ
Orders, &c.

II. CELLULARES.

CLASS 3. Acotyledoneæ.

Sub-class 1. Foliaceæ
Orders, &c.

Sub-class 2. Aphyllæ.
Orders, &c.

Examination of Species arranged according
to the Natural System.

Ex. gr. THALAMIFLORÆ.

ORDER 1. Ranunculaceæ.

Tribe 1. Clematideæ.

 Gen. 1. Clematis.

  Sect. 1. Flammula
   § 1. Species (1—21)
   § 2. ——— (22—44)
   § 3. ——— (45—54)
   § 4. ——— (55—67)
   § 5. ——— (68—71)

  Sect. 2. Viticella
   Species (72—76)

  Sect. 3. Cheiropsis
   Species (77—81)

  Sect. 4. Atragene
   Species (83—86)

 † Clematides non satis notæ
   Species (87—90)

 Gen. 2. Naravelia
   Species 1.

Tribe 2. Anemoneæ
 Gen. (3—10) &c.

Tribe 3. Ranunculeæ

    Gen. 11. Myosurus
        Species (1—2)

    Gen. 12. Ceratocephalus
        Species (1—2)

    Gen. 13. Ranunculus
      Sect. 1. Batrᵃchium
        Species (1—3)

      Sect. 2. Ranunculastrum
        Species (4—28)

      Sect. 3. Thora
        Species (29—32)

      Sect. 4. Hecatonia
        § 1.   Species (33—44)
        § 2.   ———— (45—48)
        § 3.   ———— (49—62)
        § 4.   ———— (63—135)

      Sect. 5. Echinella
        § 1.   Species (136—145)
        § 2.   ———— (146—149)

    † Ranunculi non satis noti
        § 1.   Species (150—152)
        § 2.   ———— (153—159)

    Gen. 14. Ficaria
        Species (1—2)

Tribes (4—5) &c.
    ORDERS (2—10) &c.

ORDER 11. Cruciferæ
    Sub-order 1. Pleurorhizeæ

Tribes (1—6) &c.
    Sub-order 2. Notorhizeæ

Tribes (7—11) &c.
    Sub-order 3. Orthoploceæ

Tribes (12—16) &c.
    Sub-order 4. Spirolobeæ

Tribes (17—18) &c.
    Sub-order 5. Diplecolobeæ

Tribes (19—21.)

  † Cruciferæ inquirendæ

  ORDERS (12—54) &c.

ON THE STUDY OF CRYPTOGAMIC PLANTS.

Ferns—frond—involucre—capsule.

Mosses—veil (calyptra)—bristle (seta)—capsule—lid (operculum)—peristome—teeth—columella—apophysis—

SYLLABUS

OF

LECTURES ON BOTANY,

WITH AN

APPENDIX,

CONTAINING

COPIOUS DEMONSTRATIONS OF FOURTEEN COMMON
PLANTS FOR THE ILLUSTRATION OF TERMS.

BY THE

REV. J. S. HENSLOW, A.M.,
PROFESSOR OF BOTANY.

CAMBRIDGE:
PRINTED AT THE UNIVERSITY PRESS.

1853.

SYLLABUS.

I. STRUCTURAL BOTANY.

External Organization.

I. PLANT: axis of Vegetation; Appendages. Organs.

    1. CONSERVATIVE ORGANS.

Descending Series.

II. ROOT : Tap, fibre, fibril, spongiole.
          —   Neck.   —

Ascending Series.

III. STEM : nodes, internodes. Branch. Buds.
      (Herb, Undershrub, Shrub, Tree.)
      Ærial. Definite, Indefinite. Runner, Sucker.
      Subterranean (bitten). Corm, Bulb (scaly, lami-
      nated). Tuber. Rhizome (some *epigean*).

IV. LEAF : simple, compound, decompound, supra-
      decompound. Leaflets. Petiole (general,
      partial), limb (base, margin, apex). Veins,
      ribs.
      Phillode. Pitcher.
      Stipules and stipels.
      Vernation (folded, rolled).

                                1—2

2. REPRODUCTIVE ORGANS.

V. INFLORESCENCE. Peduncle, Pedicels, Bract
(Spathe, Involucre), Bracteoles.
General-receptacle. Flowers.
Indefinite (axillary) or Centripetal.

	Simple.		Compound.
	Spike (spikelet)		
	— Catkin, Spadix.		
	Raceme	Panicle, Thyrse	
	Corymb		
	Umbel (umbellule)	General Umbel	
	Head		

Definite (terminal) or Centrifugal.
Cyme. Glomerule.
Mixed.

VI. FLOWER. Floral-whorls; (complete or incom-
plete; each in one or more series).
Floral receptacle (Torus. Disk).
Æstivation (valvate, imbricate).
Perianth (leaves).
Calyx (sepals) }(claw, tube, } regular,
Corolla (petals)∫throat, limb)∫irregular.
Andrœcium. Stamens (Filament, Anther with
Pollen)...Androphorus.
Gynœcium. Pistil (carpels). Ovary (with
ovules), Style, stigma...Gynophorus.

VII. INFRUCTESCENCE (modified condition of the parts
of Inflorescence), generally regarded as "com-
pound-fruit" in Coniferæ (*Cone, Galbulus*),
Fig (*Syconus*), Mulberry (*Sorosis*), &c.

VIII. Fruit (modified condition of ovary and ovules).

    Exuvies (accessory appendages).

    Pericarp (cells, dissepiments, valves, placenta, funicular-chord).

    Apocarpous. Achene, utricle, follicle, legume (*pod*), drupe (drupel), simple berry, &c.

    Syncarpous. Capsule, pod (*siliqua*), cremocarp, nut, key (*samara*), gourd (*pepo*), berry, pome, &c.

    Aggregate. Æterio.

IX. Seed (axis, base, apex).

    Integument,(Testa and Tegmen),Nucleus (base and apex), hile, micropyle, raphe, chalaze.

    Arillus, Arillode, Caruncula, Strophiola.

    Albumen.

    Embryo. Radicle, plumule, cotyledon, tigellum.

Appendages (*metamorphic*) to various parts.

    Armature. Thorn, Spine, Prickle.

    Supports. Tendril, Claw.

***Internal Organization, and Elementary Tissues.*

Elementary Organs, of Membrane (*Cellulose*).

The Cell; usually from .05 to .01 line in diameter (also .1 to .001), cylindrically elongated, to .3 to 2.6 lines: (also to many inches). Spherical and ellipsoidal to polygonal; sometimes branched.

Cellular tissue (*parenchyma*).

Vessels—Ducts, Spiral Vessels (*Trach#ae*).

Vascular tissue. Fibrous bundles (closters and vessels).

Origin of dotted, annular, scalariform, reticulate, and spiral cells and vessels, in deposition of secondary internal layers.

Intercellular passages. Lacunæ, Receptacles, Laticiferous channels.

Organic constituents, *non-nitrogenous*, approximating to $C_{12} O_{10} H_{10}$.

Ex. gr. Starch, Dextrine, Sugar, &c.

  nitrogenous, (Protein compounds) approximating to $C_{48} O_{14} H_{36} N_6$.

Ex. gr. Albumen, Fibrine, Caseine, Gluten, &c.

Acids, Resins, Gums, Oils, Wax, &c.

Chlorophyll, Alcaloids, &c.

Elements of Inorganic constituents, adventitious in plants, found in various combinations; viz.,

Oxygen, Carbon, Hydrogen, Nitrogen, Sulphur, Phosphorus, Chlorine, Iodine, Bromine, Fluorine, Potassium, Sodium, Calcium, Magnesium, Aluminum, Silicium, Iron, Manganese.

Minute crystals (Raphides) ; some in Biforines.

Epidermis and Cuticle.

Stomates, 12,000 (*Iris*) 120,000 (*Lilac*) on sq. inch.

Pubescence.   Hair, bristle, sting, gland.

Exogenous structure of Dicotyledons.

$\left\{\begin{array}{l}\text{Central} \\ \quad \text{system.}\end{array}\right.\left\{\begin{array}{l}\text{Pith.} \\ \text{Medullary sheath.} \\ \text{Woody layers.}\end{array}\right.\left\{\begin{array}{l}\text{heart-wood}(duramen). \\ \text{sap-wood } (alburnum).\end{array}\right.$

Central
 system.
  Pith.
  Medullary sheath.
  Woody layers. heart-wood(*duramen*).
        sap-wood (*alburnum*).

Cortical
 system.
  Cortical layers. endophlæum (*liber*).
        mesophlæum.
        epiphlæum.
  Epidermis.

Medullary Rays (*silver grain*).

Modifications in Stems and Roots, in Herbs and Trees.

Computation of ages of Dicotyledonous Trees.

Observe, anomalous distribution of woody fibres in Bryonia, definite in Cycadeæ ; disks (glandular ?) on tissue of Coniferæ.

Endogenous structure of Monocotyledons.

Scattered indefinite interlacing fibro-vascular bundles in trunks of Palms; axillary buds commonly dormant.

Modifications in herbaceous and arborescent Monocotyledons.

Approximation to the ages of Monocotyledons.

Observe, Ruscus aculeatus, the only British Monocotyledonous shrub; aphyllous. Sub-exogenous structure of rhizome of Tamus communis.

Acrogenous structure of Acotyledons.

General absence of vascular tissue in the lower groups. Definite fibro-vascular bundles in the trunks of Tree-Ferns.

Fragments of fossil wood referred, by internal organization, to their Class, and some to their Order.

General anatomy of foliaceous and floral appendages.

Preparation of Skeleton Leaves.

Articulation of organs to their support, scar left by their separation.

* * * *Glossology, (some of the more important terms grouped for ready reference).*

9

A. NUMBER. Cohesion. Abortion.

Latin.   *Greek.*

stem, &c. { -angular.	E(x)-	0 A(n)-	-œcious (1, 2 3)} *species.*
-lateral.	Uni-	1 Mon(o)-	-gonous } *stem, &c.*
branch,	Bi-	2 Di-	-clinous (1, 2) } *flower.*
&c. { -chotomous.	Tri-	3 Tri-	

{ -foliate.   Quadr(i)-   4 Tetr(a)-   -chlamydeous
 -farious.   Quinqu(e)-   5 Pent(a)-   (0, 1, 2)
 -lobed   -phyllous } *perianth.*
 -fid   Sex-   6 Hex(a)-   -merous (2, &c.)
 -partite }   Sept(em)-   7 Hept(a)-   -sepalous } *calyx.*
 -sect   Oct(o)-   8 Oct(a)-   -petalous } *corolla.*

leaf, &c. { -pinnate { geminate / binate }   Non(o)- 9 Enne(a)- -androus
 ternate   Dec(em)- 10 Dec(a)-   -dynamous (2, 4)
 pectinate. Undec(im)- 11 Endec(a)-   -adelphous (1, 2, …many) } *stamens.*
 -digitate / palmate } (2, 3) Duodec(im)- 12 Dodec(a)-
 -nerved / ribbed }   Vigint(i)- 20 Icos(a)- -gynous } *styles.*
    Mult(i)- Many. Poly-
    Pauc(i)- Few. Oligo(s)- -spermous } *seed.*

leaflets { -jugate   Semi(i-)   Half. Hemi- -cotyledonous
inflores: { -bracteate   (0, 1, 2…many) } *embryo.*

flower { -floral. / -sexual (1, 2).
anther { -locular.
pericarp, &c. { -valvular.

B.   MAGNITUDE. Absolute and relative, approximate (*sub-*), obsolete. Hair $\frac{1}{144}$, Line $\frac{1}{12}$, Nail $\frac{1}{2}$, Palm 3, Span 9, &c. inches. Unequal, oblique, dimidiate, &c.

C.   INSERTION. *Position,* radical, cauline, rameal, epiphyllous, axillary, dorsal, lateral, marginal, apical, basilar, terminal, &c.
Attachment, sessile, stipitate, petiolate, peduncu-

late, peltate, perfoliate, adnate, decurrent, amplexicaul, sheathing; articulate, versatile, &c.

Adhesion, superior, inferior (*calyx, ovary*).

hypogynous, perigynous, epigynous (*corolla, stamens*).

adnate, decurrent.

D. ARRANGEMENT, alternate, opposite, ternate, verticillate (whorled), stellate, distichous, decussate, brachiate, secund, fascicled, squarrose, cæspitose, radiant, continuous, interrupted, &c.

Valvate, imbricate, twisted, plaited, quincunxial, regular, irregular, symmetrical, unsymmetrical.

Phyllotaxis, divergence, generating and secondary spirals.

Sexuality, hermaphrodite (*monoclinous*), male and female (*diclinous*), neuter. Monœcious, Diœcious, Triœcious. ·

E. DIRECTION, erect, ascending, spreading (patent), reflex, pendant, pendulous, prostrate, oblique, reclinate, resupinate, &c.

Involute, revolute, convolute, induplicate, replicate, circinate, scorpioidal, &c.

Usual relative " *inversions*" of ovule and embryo.

Ov : Orthotropous—Campylotropous—Anatropous.

Em : Antitropous—Amphitropous—Homotropous—Heterotropous.

Radicle ... from, ... to, Hile   indeterminate.

F. SUBSTANCE, scarious, membranaceous, chartaceous, coriaceous, crustaceous, corneous, ligneous, osseous, fleshy, tuberous, succulent, gelatinous, waxy, farinaceous, herbaceous, fibrous, fistular, &c.

Petalloid, sepaloid, foliaceous.

G. FORM, (when inverted, *Ob-*).

    Solid, globose, ellipsoid, ovoid, conical, cylin-
dric (terete), angular, prismatic, clavate, fusiform, filiform, capillary, gibbous, cochleate,
two-edged (*anceps*), capitate, carinate, channelled, compressed, depressed, &c.

    Tubular, campanulate (bell-), infundibuliform
(funnel-), hypocrateriform (salver-), rotate
(wheel-), urceolate (pitcher-), cyathiform
(cup-), labiate (lip), personate (mask), ringent (gape), anomalous, cucullate (hood),
galeate (helmet), &c.

    Plane, rounded, oval, oblong, lanceolate, linear,
ligulate (strap-), ovate, cordate (heart-), reni-
form (kidney-), auricled, lunate, subulate
(awl-), acerose (needle), spathulate (spoon),
cuneate (wedge-), ensiform (sword-), deltoid,
sagittate (arrow-), hastate (halberd-), &c.

    Apex, acute, pungent, mucronate, setose, awned,
rostrate, caudate, cirrhous, pointless (*muticus*),
hooked, blunt, retuse, emarginate, truncate,
præmorse, &c.

H. DIVISION, simple, branched.

    Composition, pinnate (pari-, impari-) ⎫
      pectinate, geminate (binate), ternate,⎬ (bi-,tri-,)
    digitate, palmate, (bi-, tri-).

    Incision (*marginal*), entire, repand, sinuate,
curled, crenate (bi-), dentate (bi-), serrate
(bi-), erose (gnawed), angular, runcinate,
(*laminar*), torn, incised, cut, laciniate, lobed ($\frac{1}{3}$ to $\frac{1}{2}$),
split (-fid, $\frac{1}{2}$ to $\frac{2}{3}$), divided (-partite $\frac{2}{3}$ to $\frac{3}{4}$),
cleft (-sect, $\frac{3}{4}$ to 1); panduriform, lyrate, &c.

Separation, didymous (twin), forked, stellate, articulate, septate, granular.

Dehiscence, indehiscent, longitudinal, transverse, irregular, loculicidal, septicidal, septifragal.

I. SURFACE, even, glabrous, rough, tuberculate, viscid, glaucous, pubescence (see characters of), squamose, paleaceous (chaffy), ciliate, fringed, striated, sulcate, reticulate, rugose, punctate, lacunose, alveolate.

Venation, costate (ribbed), straight, curved, parallel, convergent, divergent, reticulate.

J. COLOUR, White, Grey, Black (as neutral).

Nomenclature from Chromatometer.

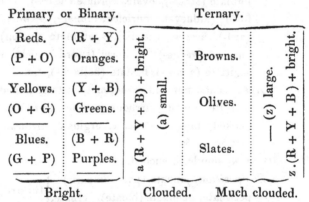

Primary or Binary.		Ternary.	
Reds.	(R + Y)		
(P + O)	Oranges.	Browns.	
Yellows.	(Y + B)		
(O + G)	Greens.	Olives.	
Blues.	(B + R)		
(G + P)	Purples.	Slates.	
Bright.		Clouded.	Much clouded.

Variegated, spotted, striped, &c.

Green (or herbaceous), as opposed to "coloured" (of any other colour).

SCENT, alliaceous, hircine, &c.

K. DURATION, annual, biennial, perennial, fugacious, caducous, deciduous, persistent (marcescent, accrescent).

II. SYSTEMATIC BOTANY.

Individual plants referred to sequence of natural " Groups."

    1 Species, (variations, varieties, races)——Hybrid.
    2 Genus, (sub-genera).
    3 Order or Family, (sub-orders, tribes, sub-tribes).
    4 Class, (sub-classes).

Genera grouped *artificially* under the Linnean System.

Flowers apparent (*phanerogamous*) ....none.

Hermaphrodite (*monoclinous*)...............unisexual   (*cryptogamous.*)
              Stamens.   (*diclinous.*)

Equality.	*Proportion.*	*Combination.*		
1 Mon-	14 Didy- ⎫	16 Mon- ⎫ -adelphia. (*filaments.*)	21 Mon- ⎫ -œcia.	24 Cryptoga- mia.
2 Di-	*Orders.*	17 Di- ⎬	22 Di- ⎭	
3 Tri-		18 Poly- ⎭		*Orders.*
4 Tetr-	Gymno- ⎱ -spermia.	*Orders.*	*Orders.*	Filices.
5 Pent-	Angio- ⎰	as classes 3	as previous	Musci.
6 Hex-		to 13.	classes.	Hepaticæ.
7 Hept- -andria.	15 Tetrady- -namia.	19 Syngene- sia. (*anthers.*)	23 Polygamia.	Lichenes.
8 Oct-	*Orders.*	*Orders.*	*Orders.*	Algæ.
9 Enne-	Siliquosa ⎱	Æqualis.	Mon- ⎫ -œcia.	Fungi.
10 Dec-	Siliculosa ⎰	Superflua.	Di- ⎬	(N.B. others
11 Dodec-		Frustranea.	Tri- ⎭	since added.)
12 Icos-		Necessaria.		
(*perig: epig :*)		Segregata.		
13 Poly-		20 Gynan- dria. (*pistil.*)		
(*hypog:*)		*Orders.*		
Orders.		as classes 1		
Mono- ⎱ -gynia.		to 13.		
Di- &c. ⎰				
Poly-				

Genera grouped *naturally* under the Jussieuan System, modified by subsequent observation.

Subordination in the value of characters derived from the reproductive organs.

1 Embryo.
2 Pistil and Stamens.
3 Seed and Pericarp.
4 Perianth and Bracts.
5 Nectaries.

Above 80,000 species is the usual estimate for *flowering* plants, probably to be reduced one third!

Selection of Orders and a few species, chiefly British, for illustrating Structural and Systematic Botany.

CLASS II DICOTYLEDONES.

Sub-class 1.   Thalamifloræ (*polypet : hypog :*)

RANUNCULACEÆ.

Refer to Demonstration I.

Ranunculus repens.   Creeping Crowfoot.

Observe, partially induplicate æstivation of Clematis.

Irregularities in perianths of several Genera.

pendulous seed of Clematis and Anemone.

follicles of Helleborus, Aquilegia, &c.

baccate pericarp of Actæa.

cohering carpels of Nigella.

PAPAVERACEÆ.

Papaver Rhæas.   Common Red Poppy.

Observe, fugacious sepals, projecting parietal placentæ *apparently* opposite radiating stigmata.   Corollæ in different species of Papaver tetra- to hexa-petalous.

CRUCIFERÆ.

Refer to Demonstration II.

Cheiranthus Cheiri.   Wall-flower.

Observe, Spuriously lateral racemes of Senebiera. Distinction of Siliqua (*a*) and Silicula (in Capsella). Septum broad (*b*) in Lunaria, narrow (*c*) in Thlaspi. Pericarp sub-indehiscent (*d*) and one-seeded in Isatis. Spurious tranverse partitions (*septa*) (*e*) in Anastatica. Lomentaceous siliqua indehiscent (*f*) in Raphanus.

Cotyledons accumbent (1) in Cheiranthus; incumbent (2) in Sisymbrium; conduplicate (3) in Raphanus; circinate (4) in Bunias; biplicate (5) in Heliophila. Hence Sub-orders and Tribes,

Sub-orders.	Embryos.		Pericarps.	
	1. Pleurorhizeæ	0 =	Siliquosæ	*a*
	2. Notorhizeæ	0 ‖	Latiseptæ . . .	*b*
	3. Orthoploceæ	0 >>	Angustiseptæ .	*c*
	4. Spirolobeæ	0 ‖‖	Nucamentaceæ	*d*
	5. Diplecolobeæ	0 ‖‖‖	Septulatæ . . .	*e*
			Lomentaceæ. .	*f*

RESEDACEÆ.

Observe, Reseda odorata, large hypogynous disk, mutifid limb at back of claw of petals, summit of capsule gaping !

VIOLACEÆ.

Refer to Demonstration III.

Viola tricolor.   Pansy Violet.

Observe, barren (irregular) and fertile (regular) states of the flowers of some species of Viola.   Caulescent and acauline species.

POLYGALACEÆ.

Polygala vulgaris.   Common Milkwort.

Observe, three bracts at base of pedicel; one posterior and two anterior (outermost) sepals equal, two lateral (interior) much larger and petaloid (wings).   Two posterior petals, and one anterior (keel), adhering to split tube formed by cohesion of filaments ;   Keel crested. Stamens diadelphous, octandrous.   Arillode on the seed.

======

Sub-class 2.   Calycifloræ (*polypet: and monopet: perig: and epig :*)

LEGUMINOSÆ.

Refer to Demonstration IV.

Pisum sativum.   Garden Pea.

Observe, Ulex europæus, aphyllous, bi-partite calyx, monadelphous stamens.   Trifolium pratense, cohering

petals (monopetalous). Legume spirally twisted in
Medicago; spuriously two-celled in Astragalus, and
many celled in Cassia; drupaceous in Dipterix; lomentaceous and polyspermous in Hippocrepis, to monosper-
mous in Onobrychis. Exostoses (*tuberous*) on roots.

British species (confined to Loteæ, Vicieæ, Hedysa-
reæ) of sub-order I. Papilionaceæ, monadelphous and
diadelphous; many exotics (in Podalyrieæ, Sophoreæ)
have stamens free. Embryo straight-homotropous in sub-
orders II. Cæsalpineæ (flowers irregular) and III. Mi-
moseæ (flowers regular).

N.B. Some species have petals and stamens sub-
hypogynous or quite so.

ROSACEÆ.
Refer to Demonstration V.
Rosa canina. Common Dog-Rose.

Observe. Monœcious Inflorescence of
    Poterium . . . . .
Monochlamydeous Perianth
    of Alchemilla . . . .   } (Sanguisorbeæ).

Calyculate calyx and geni-
    culate persistent styles of   Dryadeæ.
    Geum . . . . . .
Tailed achenes of Dryas .
AggregatedDrupels(*Etœrio*)
    of Rubus
Succulent torus of Fragaria

Follicles of Spiræa . .   Spiræeæ.
Drupe of Prunus . . .   Amygdaleæ.
Superior calyx on Pome
    (with pseudo-syncarpous } Pomeæ.
    Follicles) of Pyrus . .

2

CUCURBITACEÆ.

Bryonia dioica.　Common Red Bryony.

Observe, large tuberous root.　Solitary stipular Tendrils (twisting dextrorse and sinistrorse).　Pentandrous, flowers, triadelphous stamens.　Anthers flexuose.　Berry a baccate " Gourd" (*Pepo*, which *compare* with Cucumber, Melon, Gourd, &c.) trilocular, fleshy dissepiments. Seeds with watery pulp (*epidermis*), when dry adhering like an arillus.

GROSSULARIACEÆ.

Ribes rubrum.　Red Currant.

Observe, palmi-lobed leaf.　Inferior berry with persistent limb of calyx, unilocular, two parietal placentæ. Seeds with gelatinous testa forming the internal pulp, tegmen crustaceous.　Embryo minute, homotropous, at base of horny albumen.

UMBELLIFERÆ.

Refer to Demonstration VI.

Heracleum Sphondylium.　Common Cow-parsnep.

Observe, phyllodes of Bupleurum.　Simple Umbels of Sanicula.　Capitulum of Eryngium.

Sub-orders.

I. ORTHOSPERMEÆ, seeds plane in front (*Heracleum.*)
II. CAMPYLOSPERMEÆ, seeds furrowed in front by inflection of margins (*Chærophyllum sylvestre.*)

III. Cœlospermeæ, seeds curved in all round (*Corian-drum sativum.*)

Angelica sylvestris, primary ridges thickened (the lateral winged).

Daucus Carota, primary ridges setulose, secondary echinate.

Prangos, primary ridges winged.

Laserpitium, secondary ridges winged.

Araliaceæ.

Adoxa Moschatellina. Tuberous Moschatel.

Observe, leaves bi-ternate to ternate (*rather* ternati-sect !) Corolla tetramerous in uppermost and penta-merous in four lateral flowers of the sub-cubical head. Stamens four or five accordingly, with bi-partite fila-ments, each branch with one cell of an anther (*apparently* octandrous and decandrous).

N.B. The next six Orders will belong to sub-class 3. Corollifloræ, if more respect is paid to the monopetalous condition of the Corolla than to its non-hypogynous insertion.

Valerianaceæ.

Centranthus ruber. Red Valerian.

Observe, dichotomous cymes. Involute and nearly obsolete limb of calyx gradually expanding into a plu-mose pappus. Spurred corolla. Stamens reduced to one. Capsular pericarp, with two indistinct barren cells, and one fertile. Seed inverse, exalbuminous.

20

DIPSACEÆ.

Dipsacus sylvestris. Common Teazel.

Observe, large involucrate general receptacle of the head (*capitulum*). Flowers expanding centrifugally and centripetally from midway. Each flower furnished with a tubular involucel and paleaceous bracteole at base. Limb of calyx cuplike. Corolla quadrifid. Tetrandrous, stamens free. Utricle crowned by persistent calyx limb, and surrounded by involucel. Seed inverse, albuminous.

COMPOSITÆ.

Refer to Demonstration VII.

Senecio Jacobæa. Common Ragwort.

Observe, Sub-orders. I. TUBULIFLORÆ (*Corymbiferæ and Cynerocephalæ. Juss.*) II. LABIATIFLORÆ. III. LIGULIFLORÆ. (*Cichoraceæ J.*)

Capitulum.	Ex. gr. Species.	Lin. Orders.
Homogamous.		
(H–H–H) . . . Leontodon Taraxacum		
————— . . . Carduus nutans.		Æqualis.
Heterogamous.		
(F–H–F) . . . Bellis perennis		Superflua.
(N–H–N) . . . Centaurea Cyanus . .		Frustranea.
Monœcious.		
(F–M–F) . . . Calendula officinalis . .		Necessaria.
Diœcious.		
(M)÷(F) . . . Antennaria dioica . . .		(with Super-flua.)
Heterocephalous.		
(M)+(F)		*
Aggregate.		
((H) (H)) . . Echinops sphærocephalus		Segregata.

Pappus, stipitate and pilose in Leontodon; sessile and plumose in Carduus arvensis; absent in Bellis.

Receptacle, naked in Matricaria, chaffy in Anthemis.

Order, widely diffused and species eminently endemic.

CAMPANULACEÆ.

Observe, any Campanula in æstivation, with lines of pubescence on the style (retractile Collectors). Capsule dehiscing by pores, some opening by valves.

Jasione, with syngenesious anthers, capsule dehiscing at the summit.

ERICACEÆ.

Observe, Erica Tetralix (cross-leaved Heath), marcescent corolla, anthers dehiscing by pores near the summit and aristate below. Capsule loculicidally dehiscing, with dissepiments attached to 4 valves.

Culluna vulgaris (Ling), flowers drooping, 4 coloured bracts subtending the calyx. Capsule septicidal and septifragal, the four dissepiments adhering to the axis.

OLEACEÆ.

Observe, Fraxinus excelsior (Ash), polygamous and diœcious states of the trees. Capsule compressed, winged above (*Samara*), indehiscent, unilocular, and one-seeded by abortion.

Ligustrum vulgare (Privet), bilocular berry.

Sub-Class *3.* Corollifloræ (*Monopet: hypog:*)

CONVOLVULACEÆ.

Observe, Convolvulus Sepium (Great Bindweed), the dextrorse twining stem, two large bracts enclosing the calyx.

Cuscuta, (*Dodder*), any species, dextrorse twining leafless parasitic stem, with papillose suckers, coronal scales in the tube of the corolla, acotyledonous (*gamocotyledonous?*) filiform embryo, spirally convolute with thickened radicle, early decaying after germination.

SOLANACEÆ.

Observe, Solanum Dulcamara, corolla with plicate æstivation, anthers dehiscing by pores at the apex, bilocular berry.

Datura Stramonium, capsule bilocular, each cell divided below by an imperfect spurious dissepiment, loculicidally and septicidally quadrivalvular.

SCROPHULARIACEÆ.

Observe, Verbascum Thapsus, unequal segments of rotate corolla (approximation to Solanaceæ), pentandrous.

Rudiment of fifth stamen in Scrophularia.

Antirrhinum majus, personate corolla, stamens didynamous. Regularity restored in Peloria forms of Antirrhinum, Linaria, and Calceolaria.

Veronica Chamædrys, quadrifid rotate corolla, diandrous.

BORAGINACEÆ.

Observe, the harsh pubescence and scorpioidal inflorescence (circinate cymes) of most of the species. Achenes generally resembling those of Labiatæ, but seeds pendulous. In Lithospermum officinale nutlike stony pericarps.

LABIATÆ.

Refer to Demonstration VIII.

Lamium album. White Dead-nettle.

Observe Salvia; two inferior stamens only, and one cell abortive to each anther; elongated connective; some species with rudiments of two superior stamens. Rudiments of a fifth uppermost stamen in some monsters.

PRIMULACEÆ.

Refer to Demonstration IX.

Primula vulgaris. Common Primrose.

Observe, Embryo homotropous, in Hottonia and Samolus. Capsule a pyxis, in Anagallis. Ovary half inferior, and 5 coronal scales, in Samolus.

Forms intermediate between Primrose, Cowslip, Oxlip, and Polyanthus.

PLANTAGINACEÆ.

Observe, Plantago lanceolata. Ribwort Plantain. Calyx four-parted, imbricate. Corolla scarious. Capsule a pyxis.

24

Sub-Class 4. Monochlamydeæ {*achlamyd:* and *monochla-myd:* (to *dichlamyd: diclin:*) *hypog: perig: epig:*}

CHENOPODIACEÆ.

Observe, any Atriplex : polygamous flowers, females with accrescent perianth enclosing *compressed* utricle with *vertical* seed; others with *depressed* utricle and seed none (in males) or *horizontal.* Embryo amphitropous outside farinaceous albumen.

EUPHORBIACEÆ.
Refer to Demonstration X.
Euphorbia Helioscopia. Sun Spurge.

Observe, Mercurialis perennis; diœcious, bicoccous capsule; *compare* seeds with those of Ricinus communis, on larger scale. Buxus sempervirens; leaves hollow.

Phyllanthus; aphyllous species with dilated flattened and leaf-like branches floriferous on the margin.

N.B. Cassava and Tapioca obtained from highly poisonous Manihot !

CALLITRICHACEÆ.
Callitriche verna. Vernal Water-starwort.

Observe, sessile flowers in axils of leaves usually monœcious, between two bracteoles; male one stamen, female one pistil with two styles and quadrilocular ovary.

URTICACEÆ.
Observe, Urtica dioica, sometimes monœcious; charac-

ter of the stings; inflexed stamens in æstivation, separate elastically. Orthotropous ovule and antitropous embryo. Humulus Lupulus. Common Hop. Leaves harsh, with forked bristles. Stem, sinistrorse twining. Large bracts of the conelike female catkin. N.B. urceolate perianth early adnate to the ovary, subtended by scale-like bract (perianth *auct.*) with peculiar resiniferous glands.

Observe, Amentiferous inflorescence of three following orders.

SALICACEÆ.

Observe, any Salix. Willow. Achlamydeous flowers, each subtended by a bract, with a prominent gland on other side at the base. Comose arillus surrounding the seeds.

CUPULIFERÆ.

Observe, adnate evanescent perianth; cuplike involucre to the fruit (*glans*).

	Cells to ovary.	Ovules to Cell.
Corylus (*Hazel*)	2	2 (not 1 !)
Fagus (*Beech*)	3	1
Quercus (*Oak*)	3	2
Castanea (*Chesnut*)	± 6	1

Seeds frequently only one by abortion.

N.B. The following Order has been regarded as one of a Class apart from Dicotyledones, under the name of Gymnogenæ.

CONIFERÆ.

Observe, Leaves evergreen or deciduous (Larch), fascicled or scattered, acerose.

Female flower, (open carpellary scale) }		Infructescence (cone with naked seeds).}
Taxus (*Yew*) ...... Uni-	} ovuliferous	one seed.
Pinus (*Fir*) .... Bi-		many scales (*Strobilus.*)
Juniperus (*Juniper*) Tri-		few scales...(*Galbulus*).
Cupressus (*Cypress*) Multi-		

Fleshy cupulate arillus to seed of Taxus, and its crustaceous testa.

Fleshy baccate trifid galbulus of Juniperus.

CLASS II. MONOCOTYLEDONES.

N.B. The two following Orders have been grouped with a few others by some Botanists, as worthy of being considered a Class apart from Monocotyledones, under the name of Dictyogenæ.

TRILLIACEÆ.

Paris quadrifolia. Common Herb-Paris.

Observe, deviations from the usual quaternary arrangement of leaves and floral whorls, approximating to ternary (simply or doubly). Leaves of perianth distinguishable into sepals and petals ; ovary superior.

DIOSCOREACEÆ.

Tamus communis. Common Black Bryony.

Observe, sinistrorse twining stem, diœcious flowers, petaloid perianth and inferior ovary.

ORCHIDACEÆ.

Refer to Demonstration XI.

Orchis Morio.   Green-winged Orchis.

Observe, difference in specific gravity of the old and new tubercle. Numerous tropical Epiphytes. Peculiarities in Pollen characteristic of the four tribes which include British species.

1. Pollinia separable into numerous small masses of cohering grains.

    Ex. gr. Orchis and Ophrys .... OPHRYDEÆ.

2. Pollen granular, (grains very slightly cohering).

    Ex. gr. Listera, Neottia............NEOTTIEÆ.

3. Pollinia in definite waxy masses.

    Ex. gr. Malaxis paludosa .... MALAXIDEÆ.

4. Two lateral Anthers perfect, intermediate one barren and petaloid (*Staminodium*).

    Ex. gr. Cypripedium Calceolus...CYPRIPEDIEÆ.

IRIDACEÆ.

Observe, any Iris, equitant ensiform leaves; petaloid stigmata.

AMARYLLIDACEÆ.

Refer to Demonstration XII.

Narcissus Pseudo-narcissus.   Common Daffodil.

Observe, No crown in Galanthus.

28

Liliaceæ.

Contrast, the *inferior* perianths of this order (*Tulipa,*
Fritillaria, &c.) with the *superior* perianths of Amaryllidaceæ and Iridaceæ; and *compare* the petaloid character
and trimerous arrangement of the three orders.

Araceæ.

Refer to Demonstration XIII.

Arum maculatum.   Spotted Arum.

Observe, Richardia æthiopica, rudimentary stamens
(staminodia) about the ovaries below; and perfect
stamens covering the spadix above.

Calla palustris, flowers hermaphrodite.

Acorus Calamus, flowers monochlamydeous.

Orontium aquaticum, no spathe, and seeds exalbuminous.

Lemnaceæ.

Observe, Lemna minor.   Lesser Duckweed.   Calyp-
trate roots, proliferous cellular *frond.*   Spadix obsolete,
spathe membranous investing two stamens (male flowers),
which expand in succession, and one pistil (female flower.)

Gramineæ (*Grasses.*)

Refer to Demonstration XIV.

Bromus mollis.   Soft Brome-grass.

Observe, among British grasses,

Alopecurus pratensis, dense spike-like panicle.

Nardus stricta, glumes 0, lodiculæ 0, style 1.

Lolium perenne, glume 1.

Anthoxanthum odoratum, stamens 2; spikelets with one fertile and two barren flowers (awned scales).

Stipa pinnata, lodiculæ 3, greatly lengthened feathery kneed and twisted awn of outer pale (in the fruit.)

Among Cerealia (larger seeded grasses).

Wheat (*Triticum vulgare*), numerous varieties,
   α. æstivum.   β. hybernum.
Parentage asserted to be an Ægilops.

Spelt (*Triticum Spelta.*)

Barley (*Hordeum vulgare*), spikelets by threes, one-flowered,
   {two-rowed (*H. distichum*) 2 spikelets barren,
   {six-rowed (*H. hexastichum*) 3 spikelets fertile.

Oat (*avena sativa*).

Rye (*Secale cereale*).

Rice (*Oryza sativa*), glumes minute, pales rigid, stamens 6.

Maize (*Zea Mays*). Monœcious, panicle sometimes androgynous.

Millet (various species of *Panicum, Sorghum, Eleusine*, &c.)

Sugar Cane (*Saccharum officinarum*).

Job's-tears (stony involucrum of *Coix Lachryma*).

Bamboo (*Bambusa arundinacea*) arborescent to 60 feet, exudes Tabasheer in the joints of stem.

Ergot, diseased seeds of various grasses.

Viviparous state of certain grasses.

CYPERACEÆ (*Sedges.*)

Observe, Carex riparia, or C. paludosa, triangular solid stem (*culm*), sheath of leaves without a ligule, monœcious spiked flowers,

> *male flower*, 3 stamens subtended by a bract.
> *female flower* with urceolate perianth subtended by a bract, 3 stigmas.
> *embryo* within the albumen.

Papyrus prepared from Cyperus Papyrus.

CLASS III. ACOTYLEDONES.

GENERAL RECOGNITION OF THE ORDERS.

FILICES (*Ferns.*)

Tree-Ferns.

Circinate vernation of Fronds.

Sorus. Indusium. Theca (*Spore-case*). Annulus. Spores.

Pro-embryo state, with Antheridia and Pistillidia.

LYCOPODIACEÆ (*Club-mosses.*)

EQUISETACEÆ (*Horsetails.*)

Hygrometric elaters to the spores.

MARSILÆACEÆ (*Pepperworts.*)

Musci (*Mosses.*)

Cauline and perichætial leaves.

Antheridia and Pistillidia.

Seta. Apophysis. Theca (*spore-case*), of Sporangium and Sporangidium. Calyptra (*veil*). Operculum (*lid*).

Peristomium (*peristome*). Dentes (*teeth* 4, 8, 16, 32, 64, ∞). Epiphragma.

———

Hepaticæ (*Liverworts.*)

Spores with or without Elaters.

———

Lichenes (*Lichens.*)

Thallus, Apothecium (*shield*), Spores in Asci (*thecæ*), Gonidia.

———

Algæ.

———

Fungi.

Mycelium (*spawn.*)

———

III. PHYSIOLOGICAL BOTANY.

Mineral Kingdom } Elements {Vegetable Kingdom
  (*Inorganic.*)   } about 59. { (*Organic.*)

  (All)   (chiefly C, O, H, $\pm$ N)

Elective affinity	from	Assimilation
	Invisible	
Crystallization	to	Cell formation
	Visible	
Simple minerals	state.	Plants.

Reproductive condition of Parent Cell. Primordial
    Utricle.

Protoplasm among Cell-sap.

Appearance of Nucleus (*cytoblast*) $\pm$ Nucleoli.

Merismatic Cell formation, by septa from Primordial
    Utricle.

Free Cell formation, around aggregations of Protoplasm.

External deposition of Cellulose, as primary and second-
    ary membrane, with organic and inorganic de-
    posits.

Ex. gr. Siliceous skeleton of burnt cellular tissues, in
    glume of Oat, pubescence of Deutzia scabra, epi-
    dermis of Equisetum hyemale, lorica of Diato-
    maceæ.

Disappearance of Nuclei in many cells.

Vitality manifested in independent cells among simplest
    Plants, and in complex organs of mutually depen-
    dent cells in those more highly organized.

Stimuli affecting vital energy. Light, Heat, Moisture.

Movements depending on vitality. Sleep, Irritability.

Ex. gr. Leaves of Mimosa pudica (*Sensitive Plant*), Oxalis sensitiva, Desmodium gyrans, Dionæa muscipula, Drosera, (N.B. other fly-catching plants, Apocynum androsæmifolium, Lychnis viscosa, &c.); Stamens of Berberis, Helianthemum ; Style of Stylidium.

N.B. Exclude amount of movement depending on mechanical conditions. Ex. gr. In Dehiscence of Anthers and Pericarps ; turgescence and flaccidity from Endosmose and Exosmose in fruit of squirting Cucumber, and Balsams ; elasticity in Stamens of Nettle and Kalmia ; hygroscopicity of tissue in Mosses, Lycopodium lepidophyllum, Anastatica (*Rose of Jericho*), capsules of Mesembryanthemum, &c.

Movements of Protoplasm, ceasing in young or continued in some old cells ($\frac{1}{500}$ line per second).

Ex. gr. as observed in Caulinia; Vallisneria ; and Chara (parallel to spiral rows of chlorophyll), &c.

NUTRITION OF PLANTS.

1. Absorption, into cells; at spongioles of Roots, of fluids only. Aided by Endosmose. Modified by other physical causes.

2. Diffusion, by cells (rarely by vessels). Circulation. In exogenous stems fluids *ascend* along outer

3

layers of central system, and *descend* along inner of the cortical.

Ex. gr. Effects of ligatures and ringing, on parts above and below the wound.

Propulsion below and *attraction* above.

Force of ascent in Vine equal 26 to 38 inches of mercury, aided by Endosmose. Bleeding of cut branches.

3. Exhalation (evaporation only ?) of water at stomates in *light*, equal two thirds the absorption by spongioles.

Ex. gr. Sunflower exhales 30 oz. per diem. Drops on leaves of grasses, &c. about sunrise. Plants become dropsical in the dark.

4. Respiration, contrasted with that of animals.

Green parts in *light* absorb Carbonic acid and exhale Oxygen ; in *darkness* vice versâ.

Coloured parts (not Green) constantly exhale Carbonic acid.

Assimilation of organic (organizable) compounds chiefly effected from Carbonic Acid (CO_2), Water (OH), and Ammonia (H_3N). Whether is CO_2 or OH decomposed to set O free ?

5. Nutrition.

Fresh tissue only added by the vital activity of that last formed.

Conversion of starch to dextrine and sugar, in malting (by action of diastase)—in germination—in bud development, &c.

N.B. Compare nutritive properties of Potato with Flesh (Muscle), omitting inorganic compounds (,78 per cent.).

POTATO.		C	O	H	N
74. water	*	65,7	8,2	*
26. { 2. protein compounds . .	1,0	0,4	0,1	0,3
16. starch . . . / 8. cellulose. .	10,6	11,6	1,4	*

100, { ... }

FLESH.

		C	O	H	N
74. water	*	65,7	8,2	*
26. fibrine	13,9	5,7	1,7	4,0

100, { ... }

Blemish introduced by pruning Timber. Tree in Tree.

Grafting of allied species of Dicotyledons ; rarely succeeds in Monocotyledons. Independent growth of stock and graft

Production of organs deficient in detached stem, root, or leaf. Occasional increase of detached roots.

Organic compounds are essential to the nourishment of some plants, useful to many.

Nutrition of flowering Parasites (on Stems and Roots). (1) Assimilation imperfect when no green parts present. (2) Absorption imperfect.

Ex. gr. Viscum, Myzodendron, &c. (on stems); Thesium, Rhinanthus, Melampyrum, Euphrasia, &c. (on roots) have green leaves.

Ex. gr. Orobanche, Lathræa, Monotropa, Cynomorium, (*fungus melitensis*), &c. (on roots) ; Cuscuta, Rafflesia, &c. (on stems) have no green leaves.

Parasitic fungi imply(?) and promote disease.

Ex. gr. Uredo caries (Bunt); U. segetum (Smut); U. rubigo (Rust) precursor to Puccinia graminis (Mildew); among Wheat, &c. Ergotetia the cause of Ergot?

Botrytis infestans the *proximate* cause of " Potato Disease."

N.B.   Morbid conditions produced by attacks of Insects, &c.

Ex. gr. Numerous Galls by Cynips, &c.; Ear-cockle in Wheat by Vibrio Tritici.

Secretions, their relation to assimilated compounds.

Exhaustion of soils, and rotation of Crops.

Evolution of heat, by formation of carbonic acid, and organic compounds.

Ex. gr. By germinating seeds (raised 54^0); during fertilization of Aroideæ—in Arum maculatum (raised 27^0), in Colocasia odora (raised 50^0).

REPRODUCTION OF PLANTS.

By successive Fission of cells (into 2 or 4) indefinite, or restricted.

By production of buds (*gemmæ*) from cells, (*Musci*, &c.)

By production of buds by complex organs—some separating spontaneously.

Ex. gr. on Stems (*Lilium tigrinum, Dentaria bulbifera*). On Leaves (*Bryophyllum; Malaxis*).

Abnormal, from roots, stems, leaves, &c.

Reproduction by spores, without or by " conjugation."

Movements of Spores (*zoospores*) by ciliæ.

Reproduction by Spores in Cryptogamia with Antheridia and Pistillidia.

Reproduction by Seeds in Phanerogamia with Anthers and Pistils.

Formation and developement of the Embryo in Phanerogamia.

1. Flowering, its periodic return.

    formation of pollen grains; outer and inner membranes; fovilla (*granules*).

    grains exploded by moisture.

    vitality of some pollen preservable for a twelvemonth.

    formation of ovules; nucleus (with embryo sac) secundine, primine.

2. Fertilization.

    dispersion of pollen; frequent instrumentality of Insects.

    Ex. gr. economy in Orchis, Asclepias, Valisneria, Phœnix, Ficus.

    growth of Pollen tubes induced and maintained by stigmatic secretion.

their extension down conducting tissue (in Cereus grandiflorus to many thousand times their diameter !)

death of stigma—collapse of grains—growth of tubes for a few hours, or for many days, to reach the ovules.

formation of " germinal vesicles" (usually 3).

pollen tube penetrates micropyle and reaches embryo sac.

developement of one (rarely more) germinal vesicle into pro-embryo, and this into " suspensor" and " embryo."

Anomalous case of Cœlebogyne.

Limitation of hybridization to allied species.

Case of a Hybrid "Cytisus purpureo-laburnum," producing buds of the two parents !

3. Maturation.

developement of embryo with modifications of its integuments into ripe seed.

Dissemination of seeds; preservation of their vitality.

Germination of seeds. *Unproved* in Mummy Wheat of Egypt; or in Tomb Maize of Peru.

Botanical Geography.

Influence of Climate, regulated by Isothermal, Isotheral and Isochimenal lines.

Relative influences of Temperature, Humidity, Light, Altitude, Exposure, Soils, &c., in determining "stations" for Endemic species.

Effects of Spring frosts, and of cold by radiation.

Causes influencing dispersion of Sporadic species.

Preoccupation of surface by rapidly growing, and subsequent ejection by more robust and slow growing species.

APPENDIX.

DEMONSTRATIONS OF FOURTEEN COMMON PLANTS.

DICOTYLEDONES.

1. *Thalamifloræ.*

1. Ranunculus repens. Creeping Crowfoot.
2. Cheiranthus Cheiri. Common Wall-Flower.
3. Viola tricolor. Pansy Violet.

2. *Calycifloræ.*

4. Pisum sativum. Garden Pea.
5. Rosa canina. Common-dog Rose.
6. Heracleum Sphondylium. Common Cow-parsnep.
7. Senecio Jacobæa. Common Ragwort.

3. *Corolliflorce.*

8. Lamium album. White Dead-nettle.
9. Primula vulgaris. Common Primrose.

4. *Monochlamydeæ.*

10. Euphorbia helioscopia. Sun Spurge.

MONOCOTYLEDONES.

11. Orchis Morio. Green-winged Orchis.
12. Narcissus Pseudo-narcissus. Common Daffodil.
13. Arum maculatum. Spotted Arum.
14. Bromus mollis. Soft Brome-grass.

FORMULA

I. PLANT : caulescence, foliation, flowering.

II. ROOT : tap, fibres.
Appendages. Tubercles, exostoses.

III. STEM and BRANCHES: aërial; epigean, hypogean.
Appendages. Armature, supports, pubescence.

IV. LEAF, LEAFLETS, STIPULES:
Petiole, limb, veins, vernation.
Appendages. Armature, supports, pubescence.

V. INFLORESCENCE : peduncle, pedicels, bracts,
bracteoles, general receptacle, flowers.
Appendages. Armature, supports, pubescence.

VI. FLOWER : floral-receptacle, whorls, æstivation,
perianth, (calyx, corolla), stamens (filament,
anther, pollen), pistil (ovary, ovules, style,
stigma).
Appendages. Nectaries, awns, pubescence.

VII. Infructescence. (See Inflorescence with fruit
for flowers.)

VIII. FRUIT : Pericarp (carpels), cells, dissepiments,
valves, placenta.
Appendages. Exuvies, armature, pubescence.

IX. SEED : testa, hile, raphe, chalaze.
Albumen—Embryo (cotyledon, radicle).
Appendages. Arillus, arillode, carunculus.
Pubescence.

FOR DEMONSTRATIONS.

A. NUMBER. Cohesion, Abortion.

B. MAGNITUDE. Proportion, Extension, Abundance, Visibility.

C. INSERTION. Position, Attachment, Adhesion.

D. ARRANGEMENT. Symmetry, Regularity, Construction, Sexuality.

E. DIRECTION. Inversion.

F. SUBSTANCE. Structure, Strength, Solidity, Succulency, Vascularity.

G. FORM. Solid, Tubular, Plane.

H. DIVISION. Ramification, Composition, Incision, Separation, Perforation, Dehiscence.

I. SURFACE. Striation, Venation.

J. COLOR and SCENT.

K. DURATION. Periodic relations.

CLASS I. DICOTYLEDONES.

Sub-class 1. Thalamifloræ (*polypet. hypog.*)

DEMONSTRATION I.

RANUNCULACEÆ.—Ranunculus repens. Creeping
Crowfoot.

I. PLANT. Herbaceous, caulescent, leafy, with
dichlamydeous, bright yellow, polyandrous
flowers.

II. ROOT. G, fibrous.

III. STEM. E, erect or ascending. F, fistular. G, cylindric. H, branched. I, hairy, furrowed.
branches. A, numerous. E, upper erect, lower
prostrate creeping (runners).
N.B. Some specimens entirely prostrate.

IV. LEAF. C, radical and cauline; lower petiolate,
upper sessile, semi-amplexicaul. D, alternate. E, spreading. G, limb broadly cordate,
petiole dilated. H, radical ternati-sect (*not*
strictly compound !), cauline tripartite to
trifid. I, glabrous to pubescent. J, dark
green, lowest often with a black spot.
segments (not *leaflets !*). C, stalked. G, lower
wedge-shaped, upper lanceolate, more or less
deeply three-lobed and inciso-dentate.

V. INFLORESCENCE. C, terminal and lateral one-
flowered furrowed peduncles.

VI. FLOWER.

Torus. B, large. I, hairy.

Calyx. A, pentasepalous. C, inferior.

sepals. F, spreading (patent). G, ovate-oval,
concave. H, entire. I, pubescent outside.
J, somewhat coloured. K, deciduous.

Corolla. A, pentapetalous.

petals. B, twice size of sepals. C, hypogynous, shortly clawed. D, alternate with
sepals, æstivation imbricate. E, patent.
G, ob-cordate, obtuse. I, shining. J, golden
yellow with paler claw. K, deciduous.

*nectariferous pore covered by a notched scale,
at the base and inside of each petal.

Stamens. A, numerous (polyandrous flower).
B, half length of the petals. C, hypogynous.
E, ascending.

filament. G, filiform-subulate, compressed.

anther. C, terminal, adnate. E, erect. G,
linear-oblong. H, dehiscence lateral (not
really extrorse!).

pollen. G, globular.

Pistils (carpels). A, numerous (polygynous
flower). B, small. C, sessile. D, imbricate.

ovary. G, ovoid, compressed. H, unilocular.
I, glabrous.

style. A, one. B, very short. E, reflexed.
K, persistent.

Stigma. A, one. B, minute. H, simple.

ovule. A, one. C, basilar, sessile. E, erect, anatropous. G, ob-ovate.

VII. INFRUCTESCENCE, as Inflorescence, with peduncles elongated.

VIII. FRUIT. A, numerous aggregated Achenes (*Etœrio*). G, globular.

Pericarp (achene). C, sessile. G, obliquely ovoid, compressed, with slightly hooked beak. H, indehiscent. I, smooth, minutely pitted. J, light brown.

IX. SEED. A, one. B, fills the cell. C, basilar. E, erect.

testa. F, membranous.

albumen. B, copious. F, horny. G, ob-ovoid.

embryo. B, very minute. C, at base of albumen, homotropous. H, dicotyledonous.

DEMONSTRATION II.

CRUCIFERÆ. Cheiranthus Cheiri. Common Wall-flower.

I. PLANT. Suffrutescent, leafy, with racemose tetramerous flowers.

II. ROOT. F, woody. H, branched.

III. STEM. E, erect. F, somewhat woody. G, te-

rete, sub-angular. H, branched. I, sub-
pubescent.

branches. A, numerous. D, alternate. E,
ascending.

IV. LEAF. C, cauline, sub-petiolate or sessile.
D, alternate. E, patent. G, lanceolate,
acute. H, simple, entire, lowermost slightly
divided. I, hoary beneath.
Stipules. A, none.
*Pubescence, of hairs. E, appressed. H, bi-partite.

V. INFLORESCENCE. C, terminal. E, erect. G,
raceme, at first corymbose.
pedicels. B, shorter than flowers. G, filiform.
bracts. A, none.

VI. FLOWER. A, numerous. B, conspicuous. D,
scattered, regular, hermaphrodite. E, patent.
calyx. A, tetrasepalous. C, inferior.
sepals. D, cross-wise. E, erect. G, linear-
lanceolate, the anterior and posterior sac-
cate at the base. H, entire. I, smooth.
J, green, tinged with purple. K, deciduous.
corolla. A, tetrapetalous. B, twice length of
calyx. C, hypogynous.
petals. D, alternate with sepals, cross-wise.
E, claw erect, limb patent. G, ob-ovate
rounded, slightly apiculate or not, with sub-

linear claw. **H**, entire. **I**, smooth. **J**, yellow, to deep red-brown. **K**, deciduous.

stamens. **A**, six (hexandrous flower). **B**, tetradynamous. **C**, hypogynous. **D**, two shortest opposite anterior and posterior sepals; four longest in pairs opposite lateral sepals. **E**, erect.

filaments. **B**, shorter than claw of petals. **G**, filiform.

anther. **C**, terminal. **D**, introrse. **E**, erect. **G**, cordate-lanceolate. **H**, dehiscence longitudinal.

pollen. **G**, spherical.

Pistil. **A**, one (by cohesion of 2 carpels). **B**, length of calyx. **C**, superior. **E**, erect.

ovary. **G**, sub-linear, compressed. **H**, spuriously bi-locular.

style. **A**, one. **B**, very short. **C**, terminal, continuous. **H**, simple.

stigma. **A**, one (by 2 combining). **C**, terminal. **H**, two-lobed.

lobes. **D**, opposite placentæ. **E**, patent.

ovules. **A**, numerous. **C**, parietal. **D**, two rows in each cell, but ranging in one down the middle. **E**, horizontal, campylotropous.

*Appendages. Green hypogynous glands at the base of the shorter stamens.

VII. INFRUCTESCENCE. **G**, more elongated raceme than the Inflorescence.

VIII. Fʀᴜɪᴛ. A Pod. (*Siliqua*).

    Pericarp. B, elongated. C, pedicellate. E,
erect. G, linear compressed sub-tetragonal.
H, spuriously bilocular, bivalve, valves separating from the placentas.

    Placentas. A, two in each cell. C, margi-
nal. K, remain united by the spurious
dissepiment and form a "*replum.*"

IX. Sᴇᴇᴅ. A, numerous. C, parietal, with short
funiculus. D, one-rowed. E, pendulous.
G, ovate compressed.

    testa. F, membranous. J, brown.

    albumen. A, none.

    embryo. E, amphitropous, dicotyledonous.

    cotyledons. B, large. D, accumbent. G,
plano-convex.

    radicle. E, ascending, slightly curved; to-
wards hile.

    plumule. B, inconspicuous.

DEMONSTRATION III.

Vɪᴏʟᴀᴄᴇæ. Viola tricolor. Pansy Violet.

I. Pʟᴀɴᴛ. Herbaceous, caulescent, leafy, with irregular resupinate flowers.

II. Rᴏᴏᴛ. H, branched.

III. Sᴛᴇᴍ. B, span. E, decumbent or ascending,
somewhat zigzag. G, angular. H, scarcely
branched except below. I, downy along one
side chiefly.

4

IV. LEAF. C, cauline, petiolate. D, alternate. E,
erect. G, ovate-oblong to -lanceolate. H,
simple, broadly serrato-crenate. I, smooth,
or slightly downy.

STIPULES. A, in pairs. B, large. C, lateral, semi-
amplexicaul. G, ovate. H, lyrate-pinna-
tifid, the segments linear or tongue-shaped.

V. INFLORESCENCE.
peduncle. A, solitary. B, longer than leaves.
C, axillary. E, erect. G, semicylindric,
channelled. H, simple.
bracts. A, two. B, minute. C, towards summit
of peduncle. D, nearly opposite. G, subulate. H, simple.

VI. FLOWER. A, solitary. D, irregular, hermaphrodite. E, resupinate. H, pentamerous.
Calyx. A, pentasepalous.
sepals. C, inferior, inserted above their base,
sometimes a little cohering. D, imbricate.
E, erecto-patent. F, herbaceous. G, lanceolate acute, the base prolonged, dilated.
H, simple. I, smooth or downy. J, dark
green. K, persistent.
Corolla. A, pentapetalous. B, longer or
shorter than calyx.
petals. B, unequal. C, hypogynous, clawed.
D, alternate. E, spreading. G, two inferior rounded ; two lateral oblong ; superior (*apparently inferior*) broadly obovate-
cuneate.

*spur to superior petal. F, ascending. G, cylindrical, tubular.

Stamens. A, five (pentandrous fl.) B, equal, short. C, hypogynous. D, alternate. E, erect.

filament. B, very short.

connective. B, broad, extended beyond anthers. J, orange at the apex.

anthers. A, at first cohering (syngenesious). C, adnate. E, introrse. H, bilocular.

*spurs to two inferior entering the spur of petal. E, reclinate. G, ligulate to subclavate. J, green.

Pistil. A, one (by cohesion of three carpels ; monogynous fl.) C, superior. E, erect.

ovary. G, ovoid. H, unilocular. I, smooth. J, green.

style. A, one. C, sublateral. E, oblique. G, clavate. H, simple. K, persistent.

stigma. A, one. B, large. C, capitate. H, obliquely perforated.

placentas. A, three. C, parietal.

ovules. A, numerous. E, horizontal, anatropous.

VII. INFRUCTESCENCE. same as Inflorescence.

VIII. FRUIT. A Capsule.

Pericarp. C, sessile. F, brittle. G, ovoid subtrigonal. H, unilocular ; dehiscence loculicidal, trivalve.

    valves. E, at length horizontal. G, boat-shaped.

    placentas. A, three. C, on the valves.

    funiculus. B, very short, (obsolete).

    Exuvies of persistent Calyx.

IX. SEED. A, numerous. C, parietal along the middle of the valves. E, horizontal. G, obovoid.

    testa. F, brittle (crustaceous).

    *with a slight carunculus along the course of the raphe.

    albumen. B, copious. F, fleshy.

    embryo. B, length of albumen. C, axile. E, straight, homotropous. H, dicotyledonous.

    radicle. E, to hile.

Sub-class 2. Calycifloræ (*polypet. and monopet. perig. and epig.*)

DEMONSTRATION IV.

LEGUMINOSÆ. Pisum sativum. Garden Pea.

I. PLANT. Herbaceous, annual, with large stipules, pinnate leaves terminating in tendrils; few-flowered racemes, with rather large papilionaceous corolla.

II. ROOT. B, slender. H, branched fibres.

    Appendages. Exostoses.

III. STEM. F, weak, hollow, climbs by tendrils. H, branched. I, smooth.

IV. LEAF. C, cauline, petiolate. D, alternate.
H, impari-pinnate, but some of the terminal
leaflets metamorphosed into tendrils. I,
glabrous. J, glaucous.

   leaflets. A, two or three pair. D, mostly
opposite. G, ovate, somewhat mucronate.
H, entire, sub-repand.

   *_Tendrils_. C, from last pair or two of leaflets
and extremity of petiole. G, filiform.

   STIPULES. A, two. B, large. C, lateral. F,
foliaceous. G, ovate, sub-cordate. H, crenate towards the base. K, persistent.

V. INFLORESCENCE. C, axillary. D, alternate.
E, patent. G, raceme.

   peduncles. A, solitary. B, long.

   pedicels. A, about two or three.

   bracts. none.

VI. FLOWER. B, conspicuous. D, irregular, hermaphrodite. G, papilionaceous.

   calyx. A, monosepalous (of 5 cohering sepals).
B, 2 upper segments shortest. C, inferior.
D, irregular; æstivation imbricate, odd
sepal anterior. G, sub-campanulate. H,
quinquefid, sub-bilabiate. K, persistent.

   corolla. A, pentapetalous. C, clawed, slightly
perigynous.

   standard (*vexillum*). B, large. C, superior
(posterior) exterior. E, reflex. G, broadly
ob-cordate.

wings (*alæ*). A, two. B, shorter than standard. C, lateral. E, converging. G,
roundish.

keel (*carina*). A, of two petals, cohering by
outer edges. B, longer than wings. C,
inferior (anterior), interior. G, lunate,
compressed.

Stamens. A, ten (decandrous fl.) diadelphous,
1 free and 9 cohering. B, within the keel.
C, perigynous. E, the free one superior.

filaments. G, subulate where free.

anthers. H, bilocular, dehiscing longitudinally.

Pistil. A, one (monogynous fl.). C, superior.

ovary. G, oblong, compressed.

style. A, one. C, terminal. E, ascending.
G, compressed, subfalcate, carinated. H,
simple. I, villose above.

stigma. A, one. C, terminal. G, oblong.
H, simple.

placenta. A, one. C, along ventral suture.

ovules. A, few. C, by short funiculus. D,
in one row, but attached alternately to the
opposite valves. E, horizontal; hemi-ana-
tropous.

VII. Infructescence. In few-fruited racemes.

VIII. Fruit. A Legume. A, mostly in pairs. D,
unilateral. E, inclined.

pericarp. B, two inches. F, coriaceous.

G, oblong, swollen at ventral suture. H,
unilocular; dehiscing along ventral and
dorsal sutures, bi-valve. I, smooth.

funiculus. B, short. G, dilated (*sub-aril-*
late) above.

*Exuvies, of marcescent calyx.

IX. SEED. A, about 5 to 9. C, along ventral suture.
D, alternate on the valves. E, horizontal.
G, globose. I, smooth.

testa. F, coriaceous. I, *hile* oblong.

albumen. A, none.

embryo. B, large. E, curved-homotropous.
G, globose. H, dicotyledonous.

cotyledons. A, two. F, firm. G, plano-
convex.

radicle. E, to hile. G, conical.

DEMONSTRATION V.

ROSACEÆ. Rosa canina. Common-dog Rose.

I. PLANT. Prickly sarmentose shrub, with shining
impari-pinnate leaves, large showy flowers
and indefinite perigynous stamens.

II. ROOT. Chiefly fibres from subterranean portions
of woody suckers.

III. STEM. A, several (from suckers). B, six to
eight feet. E, straggling among bushes, or
trailing. F, stout, with large pith. G,
terete. H, branched. I, smooth.

*armature, of Prickles. B, uniform. D, scattered.
E, hooked, deflexed. F, corky inside; stout.
G, dilated at the base, sharp pointed. H,
simple.

IV. LEAVES. C, petiolate. D, alternate. H, impari-pinnate. K, deciduous.
  leaflets. A, about seven. C, sessile. D,
   opposite. E, perpendicular. F, stiff. G,
   elliptical to sub-ovate, acute. H, serrate,
   or bi-serrate. I, smooth or sub-pubescent.
  STIPULES. A, two. B, smaller than leaflets.
   C, adnate below to petiole. D, lateral. F,
   foliaceous. G, lanceolate, acute.
*armature, small prickles on the petioles, and on
ribs of the leaves.

V. INFLORESCENCE. C, definite. G, sub-corym-
bose.
  pedicles. A, one to four, or more. 1, smooth,
   or with a few gland-tipped bristles.
  bracts. D, at base of pedicels (metamorphosed stipules). G, lanceolate.

VI. FLOWER. C, pedicellate. D, regular, hermaphrodite. H, pentamerous.
  torus. C, spread over inside of the tube of
   calyx. F, fleshy. G, the disk, an annulus round the orifice of the tube. I, hairy
   inside the tube.
  calyx. A, monosepalous (5 cohering sepals).

C, inferior. D, imbricate æstivation, odd
sepal posterior. E, limb spreading. G,
tube urceolate, contracted at the throat,
segments ovate lanceolate, pointed, concave. H, limb quinque-partite, segments
pinnati-sect. I, smooth. K, tube persistent, limb deciduous.

corolla. A, pentapetalous. C, on throat of
the calyx, perigynous.

petals. B, equal. D, alternate with segments of the calyx. E, spreading and
slightly incurved. F, delicately membranous. G, ob-cordate, broadly unguiculate.
H, entire. K, deciduous.

stamens. A, indefinite (icosandrous fl.). B,
shorter than petals. C, perigynous. E,
incurved in æstivation.

filament. G, capillary.

anther. C, terminal, attached by base. D,
introrse. G, rounded-oval. H, bilocular,
dehiscing longitudinally.

Pistils (*carpels*). A, numerous, C, on lower part
of the torus (within the tube) ; slightly
stipitate. D, scattered. E, erect.

ovary. G, ovate-oblong. H, unilocular. I,
smooth.

style. A, one (to each pistil). B, slightly exserted. C, lateral. E, straight. G, filiform. H, simple. I, hairy above. K,
marcescent.

stigma. A, one. B, thickened. H, entire.

58

ovule. A, one. C, apical. E, pendulous; anatropous. G, ob-ovoid.

VII. INFRUCTESCENCE, solitary or corymbose, like the Inflorescence, but ebracteate.

VIII. FRUIT. A Hip, with numerous Achenes.

torus. E, succulent. J, red-orange.

pericarp. C, sub-stipitate. D, scattered within and towards base of calyx tube. E, erect. F, bony, with stipes fleshy, like the torus. G, ovate. H, unilocular; indehiscent. I, comose on side opposite the style (back).

*appendages. Baccate tube of calyx adhering to the torus. Marcescent stamens, and sub-clavate mass of sub-cohering styles.

IX. SEED. A, one. C, apical. D, pendulous (inverse). G, ob-ovoid.

testa. F, membranous.

raphe and chalaze. G, conspicuous.

albumen. A, none.

embryo. D, straight, homotropal. H, dicotyledonous.

cotyledons. B, large. F, fleshy. G, plano-convex.

radicle. B, small. C, superior. E, towards the hilum.

DEMONSTRATION VI.

Umbelliferæ.   Heracleum Sphondylium.   Common
Cow-parsnep.

I.   Plant, herbaceous, perennial, caulescent, leafy ;
with small pentamerous flowers in compound
umbels.

II. Root. F, somewhat fleshy. G, tap.

III. Stem. E, erect. F, fistular, with solid nodes.
G, terete. H, branched. I, furrowed,
rough with spreading hair.

IV. Leaf. C, radical and cauline, petiolate,
broadly sheathing. D, alternate. G, cordate. H, considered as impari-pinnate (but
only spuriously compound !) to ternate.
leaflets. A, two pair. C, sub-petiolate. D,
opposite. G, broadly cordate. H, variously pinnatifid cut and serrate. I, downy
beneath, veiny. J, paler beneath.
petiole. B, large in the lower leaves, ribbed ;
in the upper sheathing, the limb becoming
much diminished.

V.   Inflorescence. C, terminal to stem and axillary branches. G, compound Umbel. H,
many rayed.
pedicels (in the Umbellules). G, angular. I,
downy on one side.
bracts (in general involucre). A, few or none.

F, membranous. G, lanceolate, acute. I, ciliate at edges. K, early deciduous.

bracteoles (in partial involucres). A, several. G, lanceolate acute.

VI. FLOWER. A, numerous. D, hermaphrodite.
calyx. A, monosepalous (by cohesion of 5 sepals). C, inferior. H, limb minutely five-toothed.
corolla. A, pentapetalous.
petals. B, unequal, the outer largest, radiant. C, surrounding an epigynous disk. G, obcordate, with inflexed apex. I, white to reddish. K, deciduous.
stamens. A, five (pentandrous fl.). C, epigynous. D, alternate with the petals.
filament. B, short. G, filiform.
anther. D, introrse. G, ovate, sub-didymous. H, bilocular, dehiscing longitudinally.
Pistil. A, one (of 2 cohering carpels), often abortive in the middle flowers of the umbel.
ovary. C, inferior. H, bilocular.
styles. A, two (digynous fl.). C, terminal.
stigma. C, terminal.
ovules. A, one in each cell. C, apical. E, pendulous ; anatropous.
*appendage, Epigynous Disk (*stylopodium*).

VII. INFRUCTESCENCE. Ebracteate compound Umbel, reduced from the state of the Inflorescence by abortion of some of the middle flowers.

VIII.   Fruit, a Cremocarp.

    floral axis (Column). B, length of the cremocarp. C, between the carpels. F, slender. G, filiform. H, bi-partite.

    pericarp. A, of two cohering carpels (*mericarps*). C, suspended at summit of column. G, round, depressed, discoid, margins dilated (winged). H, emarginate, dehiscence septicidal (along *Commissure*), but the carpels indehiscent.

    ridges. A, primary 10, secondary none. C, 3 dorsal approximate, 2 lateral distant and close to margin. G, very slender.

    vittæ. A, solitary in the interstices, twin on the commissure. B, shorter than the carpels. G. sub-clavate. J, dark brown.

    appendages, divergent persistent styles.

IX.   SEED. A, one in each carpel. C, apical. E. pendulous.

    testa. C, appressed to pericarp.

    albumen. B, copious. E, straight. F, sub-corneous. G, sub-plano-convex.

    embryo. B, small. C, in the base (uppermost part) of albumen. E, straight; homotropous. H, dicotyledonous.

    cotyledons. B, somewhat unequal. G, oblong.

    radicle. C, superior. E, towards hile.

DEMONSTRATION VII.

COMPOSITÆ. Senecio Jacobæa. Common Ragwort.

I. PLANT. Herbaceous, perennial, branched, with furrowed stems, doubly pinnatifid leaves, and heads of yellow florets.

II. ROOT. Fibres from a somewhat fleshy rhizome.

III. STEM.

> *subterranean rhizome.* C, at base of aerial portion. G, somewhat fleshy, præmorse.
>
> *aerial.* A, one. E, erect. G, terete. H, branched. I, smooth furrowed.

IV. LEAF. C, cauline, lower petiolate, upper semi-amplexicaul. D, alternate. G, ob-ovate-oblong, sub-lyrate, segments oblong. H, pinnatifid, the segments pinnatifid and variously notched and toothed. I, smooth (glabrous).

> *stipules.* A, none.

V. INFLORESCENCE. A, Head (*Capitulum*). C, terminal. D, corymbose. G, heterogamous (sometimes homogamous.)

> *peduncles.* B, longer than heads. E, erect. I, cottony.
>
> *involucrum.* G, sub-hemispherical.
>
> *bracts (phyllaries).* D, two-rowed.
>
> > *innermost.* A, numerous. B, equal. D, contiguous. E, parallel. G, linear. J, tipped with black.
> >
> > *outermost.* A, fewer. B, minute. D, lax.
>
> *general receptacle.* G, flat, or somewhat convex. I, naked.

VI. Flower (*floret*). A, numerous. D, capitate.
G, ligulate in the ray, tubular in the disk.

calyx. C, superior, *tube* adnate. G, *limb*
pappose.

> *pappus.* B, length of tube of corolla. C,
> sessile. D, many rowed. H, pilose.
> I, scarcely toothed. K, persistent.

corolla. A, monopetalous (by cohesion of 5
sepals). C, epigynous. J, golden-yellow.

—(*disk florets*). G, tubular, funnel-shaped
(infundibuliform). H, 5-toothed.

—(*ray florets*). E, spreading, at length re-
curved. G, ligulate, linear-oblong. H,
3-toothed.

N. B. Ray florets sometimes wanting.

stamens (in *disk florets*). A, five (pentandrous
fl.). B, exserted. C, on tube of corolla.
D, alternate with teeth of corolla. E, erect.

—— (in *ray florets*). A, none.

filaments. B, short. G, filiform-clavate.

anthers. A, cohering (syngenesious). C,
terminal. G, linear. H, bilocular; dehiscing longitudinally.

pollen. G, spherical. I, echinulate.

Pistil. A, one.

ovary. C, inferior. G, ob-ovately sub-cylindric. I, rough.

style. A, one (monogynous fl.). B, slightly
exserted. C, articulate to summit of ovary.
E, erect. G, filiform. H, bifid at summit.
I, collectors at the apices of the branches.

> *stigmas.* A, two. C, along upper (inner) surface of the branches of the style.
>
> *placenta.* C, basilary.
>
> *ovule.* A, one. E, erect; anatropous.

VII. INFRUCTESCENCE, corymbose Heads, as in Inflorescence.

VIII. FRUIT. An inferior Achene.

> *Pericarp.* C, articulate to the general receptacle. E, chartaceous, brittle. G, obovately sub-cylindrical. H, unilocular; indehiscent. I, in the disk hairy or silky, in the ray smooth.
>
> *funiculus.* B, very short.
>
> **Appendage,* the pappus enlarged. D, manyrowed. G, pilose. K, of *ray* deciduous.

IX. SEED. A, one. C, basilary; sessile. E, erect. G, cylindric ob-conical.

> *albumen.* A, none.
>
> *embryo.* E, straight, homotropous. H, dicotyledonous.
>
> *cotyledons.* G, flat.
>
> *radicle.* B, short. E, to the hile.
>
> *plumule.* B, inconspicuous.

Sub-class 3. Corolliflorae (*monopet. hypog.*)

DEMONSTRATION VIII.

LABIATÆ. Lamium album. White Dead-nettle.

I. PLANT. Herbaceous, perennial, with quadrangular stems, opposite exstipulate leaves, bilabiate flowers disposed in axillary dichotomous cymes.

II. Root. C, fibrous, from the nodes of subterranean rhizomes.

III. Stem. A, several. C, subterranean (rhizomes) and aërial. E, rhizomes creeping; aërial erect, or ascending. F, fistular, solid at the nodes. G, quadrangular. H, simple (the branches being accounted as stems!). I, slighty pubescent.

IV. Leaf. C, radical and cauline; lowest petiolate, uppermost nearly sessile. D, opposite and decussate. E, horizontal. G, rounded-cordate, acuminate. H, simple, strongly but unequally serrate. I, rugose, sub-hirsute; venation pinnate, reticulate.
Stipules. A, none.

V. Inflorescence. C, axillary. D, forming Verticillasters. G, Cyme. H, dichotomous.
peduncle and pedicels. B, very short.
bracts. A, one to each flower. B, very minute (bracteole). C, close to calyx. G, linear-subulate.

VI. Flower. A, about ten to a cyme. D, crowded, hermaphrodite. G, irregular.
calyx. A, monosepalous (by cohesion of five sepals). C, inferior. D, somewhat irregular. G, tubular, campanulate, dilated upwards, with ten obscure angles. H, five-cleft. K, persistent.
calyx-teeth. B, nearly equal. D, upper one

5

apart. E, spreading. G, subulate.

corolla. A, monopetalous (by cohesion of 5 petals). B, twice length of calyx. C, hypogynous. G, irregular, bilabiate, ringent. *Throat* inflated, compressed, gibbous. *Upper lip* galeate. H, *Lower lip* trifid ; *middle lobe* large, obcordate, and contracted at the base ; *lateral lobes* acute.

*Ring (annulus) of hair inside the tube and near the base. E, oblique.

stamens. A, four (tetandrous fl.). B, two inferior (anterior) longest (didynamous). C, on the tube of corolla. E, ascending.

filaments. G, subulate. I, villose towards summit.

anthers. C, attached at back. D, approach by pairs. E, incumbent, cells at length divaricate. G, oblong. H, cells dehiscing longitudinally in continuous line through divaricate cells. I, hairy. K, blackish purple.

pollen. G, ovate-globose when moist, ellipsoidal dry. I, dry, three-striated. J, yellow.

Pistil. A, one (by cohesion of two or four carpels ?) C, somewhat stipitate (by gynophorus).

*Gynophorus. F, fleshy, sub-glandular. G, annular. H, four lobes alternating with the lobes of the ovary.

ovary. H, quadri-partite, divisions (lobes) truncate.

style. A, one (monogynous fl.). B, length of

stamens. C, in the middle from between the lobes of the ovary. G, filiform. H, apex bifid, divisions nearly equal.

stigmas. A, two. G, on the summits of the segments of the style.

ovules. A, one in each cell. C, basilar. E, ascending : anatropous.

VII. INFRUCTESCENCE. In crowded Verticillasters, from axillary dichotomous cymes.

VIII. FRUIT. Four sub-cohering Achenes (by subdivision of ovary).

pericarp. G, acutely triquetrous, convex on one side, obliquely truncate. H, indehiscent.

*Exuvies, of persistent (scarcely accrescent) calyx.

IX. SEED. A, one. C, basilar. E, erect.

albumen. B, extremely slight or none (exalbuminous seed).

embryo. E, homotropous. H, dicotyledonous.

cotyledons. A, two. B, large. E, straight.

radicle. C, inferior. E, to the hile.

DEMONSTRATION IX.

PRIMULACEÆ. Primula vulgaris. Common Primrose.

I. PLANT. Herbaceous, perennial, acauline, large simple leaves, single-flowered scapes or umbels of showy monopetalous flowers.

5—2

II. ROOT. B, stout fibres. C, from under side of
subterranean rhizome.

III. STEM. C, subterranean rhizome. E, oblique.
F, fleshy. G, præmorse. H, simple. I,
scaly from persistent bases of old leaves.
J, reddish.

IV. LEAF. C, radical, sub-petiolate, semi-amplexicaul. D, crowded. E, spreading. G, ob-
ovate-oblong, tapering at base. H, simple,
unequally crenate-dentate. I, rugose,
somewhat downy; venation reticulate.
STIPULES. A, none.

V. INFLORESCENCE. C, radicle (Scape). E, erect.
G, simple umbel, or usually many single
flowers (from contracted peduncle leaving
only pedicels apparent).
bracts. A, one, at base of each pedicel. B,
small. E, erect. G, subulate. H, simple.

VI. FLOWER. D, regular, hermaphrodite. E, erect.
H, pentamerous.
Calyx. A, monosepalous (by cohesion of 5
sepals). C, inferior. D, æstivation valvate.
E, erect. G, prismatic, 5-angular, sub-
inflated. H, quinquefid. K, persistent.
Corolla. A, monopetalous (by cohesion of
5 petals). B, tube longer than calyx.
C, hypogynous. D, lobes alternate with
teeth of calyx; æstivation imbricate.

E, limb spreading. G, salver-shaped (hy-pocrateriform,) throat dilated. H, limb five-lobed, lobes ob-cordate. J, sulphur yellow, with darker spot in the middle. K, marcescent.

Stamens. A, five (pentandrous fl.). B, included. C, inserted at top of the tube of corolla (*base* of throat). D, opposite the lobes of corolla. F, erect.

filament. B, short. G, filiform.

anther. C, cells parallel, introrse, attached at back above the base. E, erect, converg-ing. G, oblong, pointed. H, bilocular, dehiscing longitudinally. J, yellow.

pollen. G, globose when moist, angular when dry.

Pistil. A, one (by cohesion of 5 carpels). C, superior. E, erect.

ovary. G, globose. H, unilocular.

style. A, one (monogynous fl.). B, long as tube of corolla. C, terminal. E, erect. G, filiform. H, simple.

stigma. A, one. G, capitate. H, simple.

placenta. A, one. B, large. C, central. G, globose.

ovules. A, numerous. C, peltate. E, semi-anatropous? G, plano-convex.

VII. INFRUCTESCENCE, as Inflorescence.

VIII. FRUIT. A Capsule.

Pericarp. F, chartaceous. G, ovate-cylin-

drical. H, dehiscing longitudinally at apex
into 5 to 10 teeth. J, brown.

*Exuvies, an accrescent calyx. B, as long as
the capsule.

IX. SEED. A, numerous. C, sessile, peltate. G,
plano-convex, angular. I, testa punctate-
rugose.

albumen. B, copious. F, fleshy to horny.

embryo. B, small. C, in axis of albumen.
E, straight, heterotropous (transverse or
parallel to hile). G, sub-cylindric. H, dicotyledonous.

cotyledons. B, short.

radicle. E, indeterminate.

Sub-Class 4. Monochlamydeæ {achlamyd. and mono-
chlamyd. (to dichlamyd. diclin.) hypog.
perig. epig.}

DEMONSTRATION X.

EUPHORBIACEÆ. Euphorbia helioscopia. Sun Spurge.

I. PLANT, herbaceous, annual, foliaceous, smooth,
lactescent; small heads of one female and
several male incomplete flowers in calyciform
involucres (resembling monochlamydeous
hermaphrodite flowers).

II. ROOT. B, slender. H, branched fibres.

III. STEM. A, solitary. E, erect. G, terete. H,
branched.

branches. D, alternate below, umbellate above.
H, bi-tri-chotomous.

IV. LEAF. B, uppermost largest. C, cauline,
sessile. D, alternate. E, patent. G, ob-
ovate-cuneate. H, simple; serrated towards the apex. I, smooth. J, full grass
green.

STIPULES. A, none.

V. INFLORESCENCE. C, terminal. D, on five di-
to tri-chotomous branches umbellately disposed. G, small head (*Glomerule*) in calyciform involucre.

pedicel. B, of the female longer than of the
males.

bracts. C, on and below umbellate branches.
D, involucrate. F, leaflike. G, ob-ovate,
cuneate. H, serrated towards apex.

involucre (*Anthodium*). A, monophyllous.
F, calyciform. G, campanulate-turbinate.
H, quadrifid.

(its *laciniæ*). F, membranous. H, ciliately
cut.

*Appendages. Glands on involucre. A, four.
B, large. C, outside near the summit,
peltate. D, alternate with the lacineæ.
F, fleshy. G, roundish. H, entire. I,
shining. J, yellow green.

bracteoles. A, numerous within the involucre. B, very small. C, at base of the pedicels. H, ciliato-lacerate.

72

VI. FLOWER. D, diclinous (and monœcious), achlamydeous.

   Males. A, about 12. B, very small. C, pedicellate.

    stamen. A, one (monandrous fl.).

    filament. B, thick as pedicel. C, articu-
lated to pedicel.

    anther. C, terminal. G, cells globose. H,
bilocular, didymous.

    pollen. G, globose. I, smooth.

   Female. A, one. B, at length exserted. C, pedicellate. D, central. E, at length deflexed.

    pistil. A, one (three cohering carpels). C,
sessile.

    ovary. G, sub-globose. H, three-lobed,
trilocular.

    styles. A, three, slightly cohering (monogy-
nous fl.) at base. C, terminal. E, spreading. H, bifid. K, persistent.

    stigmas. A, six (or three bifid?) G, obtuse.

    ovules. A, one in each cell. C, apical. E,
anatropous.

VII. INFRUCTESCENCE. A, solitary.

VIII. FRUIT, a Capsule (*Regma*).

   Pericarp. G, globular. H, three-lobed, tri-
coccous, dehiscence septicidal and separating
from a central receptacle; and the carpels
(cocci) elastically bivalve.

IX. SEED. A, one in each carpel. C, apical. E,
pendulous. G, ellipsoidal.

testa. F, crustaceous. I, reticulate, pitted.

*_Appendage._ A fleshy arillode near the hile.

tegmen. F, membranous.

chalaze. B, distinct. C, opposite the hile.

raphe. B, distinct.

albumen. B, copious. F, fleshy and oily.

embryo. B, length of albumen. C, axile. E, straight, homotropous. H, dicotyledonous.

cotyledons. B, very thin. F, leaf-like. G, flat, oval.

radicle. C, superior. E, towards hile.

CLASS II. MONOCOTYLEDONES.

DEMONSTRATION XI.

ORCHIDACEÆ. Orchis Morio. Green-winged Orchis.

I. PLANT, herbaceous, perennial, with leafy scape-like stem, terminating in spike of irregular purple flowers.

II. ROOT, of fibres and tubercles (*tuberous fibres*).
fibres. F, fleshy. H, simple.
tubercles. A, two. B, size of hazel-nut. F, fleshy. G, ob-ovoid to globose. H, simple. K, annual.

III. STEM. E, erect. G, terete.

IV. LEAF. C, cauline, sheathing. D, alternate. E, lower spreading, upper erect. G, linear-

lanceolate. H, simple, entire. I, glabrous,
venation parallel. J, paler beneath.
STIPULES. A, none.

V. INFLORESCENCE. C, terminal. G, a lax spike.
bracts. A, one subtending each flower. B,
long as ovary. E, erect. F, membranous.
G, lanceolate. H, simple. I, greenish-
purple. K, persistent.

VI. FLOWER. A, four to ten. D, normal position
reversed, rather lax, irregular, hermaphrodite, gynandrous.
perianth. C, superior. D, leaves (*rather,*
segments!) in two rows. G, ringent. J,
all coloured.
outer row (*sepals*). A, three. B, equal. D,
anterior uppermost. E, converging. G,
lanceolate. I, many green ribs.
inner row (*petals*). A, three. B, two equal,
the third (*Lip*) in front larger. D, alternate with the 3 outer.
—(*two upper* do.) E, connivent, forming a
galea with uppermost one of outer row. G,
lanceolate. I, three green ribs.
—(*Lip*.) E, pendulous. C, adhering to base
of the column. G, ob-cordate. H, with 3
unequal lobes, the midde truncate, emarginate, the lateral largest, more or less
notched and irregularly crenate.
Appendage, a spur behind. B, shorter than

ovary. E, ascending. G, conical, obtuse, sub-clavate.

Central-Column. D, gynandrous. F, fleshy.

anther. A, one (monandrous fl.). C, terminal, sessile, adnate. D, cells parallel. E, erect. G, ob-ovate. H, bilocular, dehiscing longitudinally.

*staminodia. A, two. C, on either side the fertile anther. G, bluntly conical.

pollen-masses (*pollinia*). A, two. C, each with a gland (*retinaculum*) at extremity of its stalk (*caudicula*), terminating within one pouch (*bursicula*). G, clavate. H, easily separable into numerous small masses of cohering grains.

ovary. C, inferior. E, twisted, patent. G, sub-cylindric. H, unilocular.

style. C, forming a projection (*rostellum*) from its summit between the base of the anther lobes.

stigma. A, one (monogynous fl.). B, dilated. C, top of the column. E, oblique. G, concave. I, mucilaginous (viscid).

ovules. A, very numerous. B, very minute. C, to short funicular-chords, parietal. E, anatropous ?

VII. INFRUCTESCENCE, a spike of capsules.

VIII. FRUIT, a Capsule.

pericarp. F, membranous. G, ellipsoidal. H, unilocular, trivalve, valves dehiscing, from intermediate ribs, but cohering at base and apex.

placentas parietal. A, three. C, on the valves.

IX. SEEDS. A, numerous. B, very minute. C, parietal.

testa. C, loose. G, attenuated at each extremity. I, reticulate.

albumen. A, none.

embryo. F, fleshy, solid. H, monocotyledonous.

DEMONSTRATION XII.

AMARYLLIDACEÆ. Narcissus Pseudo-narcissus. Common Daffodil.

I. PLANT. Bulbous, perennial, linear radical leaves; scape bearing a single showy flower with 6-partite perianth and large bell-shaped crown.

II. ROOT. A, many fibres. D, round margin of a circular disk at base of the bulb. G, cylindric. H, simple. J, white. K, annual.

III. STEM. A, *aërial* none (acauline plant); *subterranean* within a bulb.

BULB. D, tunicated. F, fleshy, with outer coat scarious. G, sub-globose. J, blackish-brown.

IV. LEAF. A, about four. B, long as the scape. C, radicle, sheathing. D, crowded. E, erect.

F, slightly succulent. G, linear, obtuse. H, simple, very entire. I, smooth, striated, venation parallel. J, slightly glaucous.

V. INFLORESCENCE. A, solitary. C, radicle. *peduncle* (scape). E, erect. G, bluntly two-edged.

bract (spathe). A, one. B, investing flower in æstivation, afterwards one-third its length. C, at base of ovary, sheathing. F, membranous. G, ovate, concave, acute. H, entire. I, pale green. K, persistent.

VI. FLOWER. A, solitary. B, large. D, regular, hermaphrodite. E, slightly inclined, (subcernuous). G, hexamerous.

Perianth. A, single (by cohesion of six leaves). C, superior, segments alternating in inner and outer series of three each. E, limb erect, patent. F, petaloid, somewhat succulent. G, somewhat funnel-shaped (subinfundibuliform), segments lanceolate, or ovate-lanceolate, equal. H, sex-partite. I, smooth. J, pale yellow. K, marcescent.

*Crown (corona). A, one (by cohesion of abortive stamens ?). B, rather longer than perianth. C, on mouth of tube. E, erect. F, petaloid. G. bell-shaped (campanulate), the mouth expanding. H, crenate, crisped.

Stamens. A, six (hexandrous fl.). B, alternately shorter, included. C, at base of the

crown. D, alternating in two series, with
segments of perianth. E, erect.

filament. C, very slight adhesion with tube
of perianth. G, subulate.

anther. C, terminal, attached below middle
of the back. D, introrse. E, incumbent,
converging. G, linear. H, two-lobed, dehiscence longitudinal on the edges. .

pollen. G, ovoid when moist, to elliptic
when dry.

Pistil. A, one (by cohesion of three carpels).
E, erect.

ovary. C, inferior. D, carpels alternate with
inner series of stamens, i.e. are opposite
outer segments of perianth. G, sub-globose, obtusely trigonous and three-furrowed.
H, trilocular.

placenta. D, axile in each cell.

ovules. A, numerous. C, axile. D, linear.
E, horizontal, anatropal.

style. A, one (monogynous fl.). B, included.
C, terminal. E, erect. G, somewhat triangular, slender. H, simple.

stigma. A, one. C, terminal. G, obtuse.
H, somewhat three-lobed.

VII. INFRUCTESCENCE. Solitary, with scarious spathe.

VIII. FRUIT. A Capsule.

Pericarp. C, shortly pedicellate. E, inclining.
F, membranous. G, ovoid, obtusely trigonous. H, trilocular; dehiscence loculi-

cidal, trivalve, with the dissepiments in the
middle of the valves.

SEED. A, numerous. C, sessile. D, two-rowed.
E, horizontal, anatropous. G, sub-globose.
testa. I, rugose. J, black.
albumen. B, copious. F, fleshy.
embryo. B, more than half length of seed.
C, axile, and toward the base. E, homotropous, slightly curved. G, sub-cylindric.
H, monocotyledonous.
radicle. E, to the hile ; centripetal.

DEMONSTRATION XIII.

ARACEÆ.   Arum maculatum.   Spotted Arum.

I.   PLANT.   Herbaceous, perennial, tuberous, with
shining hastate radicle leaves, and large convolute spathe investing a spadix of monœcious aggregate achlamydeous flowers.

II.   ROOT.   C, fibres, from and above a tuber.   H,
simple.

III.   STEM.   C, subterranean tuber.   E, horizontal.
F, fleshy and full of starch.   G, irregularly
ovoid, præmorse.   H, simple.   I, scarred.
J, brown.

IV.   LEAF.   A, two or three.   B, large.   C, radicle petiolate, sheathing the scape.   D,
alternate.   F, erect.   G, broadly sagittate

to hastate, acute ; petiole channelled. H, simple, entire, wavy. I, glossy, strongly and palmately veined. J, dark green, often spotted with black.

V. Inflorescence. A, solitary. E, erect. H, simple.

peduncle (a Spadix), the summit of a radicle branch (scape). F, fleshy. G, clavate. I, naked above. J, yellow turning to purple. K, summit deciduous.

involucre (a Spathe). B, very large. C, sessile. D, convolute. F, foliaceous. G, oblong, narrowed above, and contracted below the middle. H, entire. J, pale green, often spotted with black outside, coloured within. K, marcescent.

VI. Flower. A, numerous. B, small. C, sessile, below the fusiform summit of the spadix. D, monœcious, grouped round the spadix, the females lowest.

males. C, sessile. D, aggregate.

Perianth. A, none (achlamydeous fl.).

stamen. A, one (monandrous fl.). B, very short.

anther. E, lobes diverging. H, bilocular, dehiscing towards the summit.

pollen. G, globose.

females. C, sessile. D, aggregate.

perianth. A, none (achlamydeous fl.).

pistil. A, one (monogynous fl.).

ovary. G, ovate. H, unilocular.

style. A, none.

stigma. C, sessile, capitate, sub-lateral. G, hemispherical, depressed.

ovules. A, few. C, parietal. D, superimposed. E, erect; orthotropous.

*Appendages, incomplete flowers of both sexes resembling each other. A, numerous. D, whorled. G, ovate and tapering at the summit.

—*staminodia* (abortive stamens). C, aggregate above the male flowers.

—*pistillidia* (abortive pistils). C, aggregate above the female flowers.

VII. INFRUCTESCENCE. C, terminal. G, spiked berries, the summit of the spadix with the male flowers having fallen.

*spathe marcescent.

VIII. FRUIT. A Berry.

pericarp. C, sessile. F, fleshy. G, ovate. H, unilocular; indehiscent. I, smooth. J, scarlet.

IX. SEEDS. A, few. C, parietal. G, sub-globose.

testa. F, sub-coriaceous, thickened at the hile.

hile. B, broad.

albumen. F, farinaceous.

embryo. C, axile, towards the summit. E,

nearly straight; antitropous. G, subclavate. H, monocotyledonous.

radicle. E, from the hile. G, obtuse.

plumule. C, somewhat denuded.

DEMONSTRATION XIV.

GRAMINEÆ. Bromus mollis. Soft Brome-grass.

I. PLANT. Annual, erect with soft pubescence,
panicle erect, rather close.

II. ROOT. G, fibrous.

III. STEM (*Culm*). A, one. B, two feet or more.
E, erect. F, fistular. G, cylindrical
(terete), swollen at the joints. H, simple.
I, smooth or slightly hairy, the joints
densely so.

IV. LEAF. C, cauline, sheathing. D, distich-
ously alternate. G, linear. H, simple.
I, nervation parallel, densely hairy.

Ligule (axillary stipule ?) B, short. G, blunt.

V. INFLORESCENCE. B, about 3 inches. D, close,
or slightly spreading. E, erect. G, a
Panicle. H, branched simply above, compound below.

branches. D, half-whorled. G, angular. I,
downy.

spikelets. A, numerous. C, pedicellate. E,

nearly erect. G, ovate-oblong, acute, rather tumid. H, many flowered.

glumes. A, two. B, unequal, nearly as long as the pales. G, elliptic, acute. I, downy, except at the base, (sometimes smooth). J, *inferior* 5-nerved; *superior* 7-nerved, the margins diaphanous.

VI.   FLOWER.   A, 7 to 10 in a spikelet.   D, closely imbricate, distichous, hermaphrodite. G, broadly elliptic, concave-depressed, the margins obtuse-angled.

Pale, outer or inferior. F, herbaceous, diaphanous at the margins. G, convex at the back, blunt. H, deeply cloven at the apex. I, downy (sometimes smooth), with 7 to 9 strong green ribs; awned below the summit.

Awn to outer pale. B, length of pale. C, a little below its summit. E, straight. F, stout. I, rough.

Pale, inner or superior. B, shorter than outer. E, membranous with green border. G, obovate, bicarinate. H, entire. I, pectinately ciliate at the edges.

Squamules (lodiculæ). A, two. B, minute, little longer than the ovary. C, hypogynous. D, collateral. E, erect. F, succulent, transparent. G, cultriform, somewhat obtuse. H, entire.

Stamens. A, three (triandrous flower). B, sub-equal. C, hypogynous. D, one anterior and two lateral.

filaments. G, filiform.

anthers. E, incumbent. G, linear, blunt. H, bilocular, and bilobed on each side, dehiscing longitudinally.

pollen. G, sub-globose. I, smooth.

Pistil. A, one (by suppression of 2 carpels?). C, superior sessile, or very slightly stipitate.

ovary. H, unilocular. I, smooth, hairy on the summit.

styles. A, two (digynous fl.). B, very short. C, sub-lateral, distant.

stigmas. A, two. C, sub-sessile. H, plumose with simple hyaline spreading hair.

ovule. A, one. E, anatropous.

VII. INFRUCTESCENCE. Contracted panicle, with spikelets articulate.

VIII. FRUIT. A Caryopsis.

pericarp. C, adnate to testa. F, thin-chartaceous. G, plano-convex. I, apex villose.

*Exuvies, the two pales.

IX. SEED. A, one.

integument (testa). A, one.

albumen. B, copious. F, farinaceous.

embryo. C, near the base on the anterior
   surface of the albumen. G, cylindric. H,
   monocotyledonous.

cotyledon. F, fleshy. G, scutelliform with
   a furrow (sulcus) in front.

radicle. G, blunt.

FOR PRIVATE DISTRIBUTION.

———

THE following pages contain Extracts from LETTERS addressed to Professor HENSLOW by C. DARWIN, Esq. They are printed for distribution among the Members of the Cambridge Philosophical Society, in consequence of the interest which has been excited by some of the Geological notices which they contain, and which were read at a Meeting of the Society on the 16th of November 1835.

The opinions here expressed must be viewed in no other light than as the first thoughts which occur to a traveller respecting what he sees, before he has had time to collate his Notes, and examine his Collections, with the attention necessary for scientific accuracy.

CAMBRIDGE,
Dec. 1, 1835.

EXTRACTS,

&c.

RIO DE JANEIRO, *May* 18, 1832.

WE started from Plymouth on the 27th December 1831—At St Jago (Cape de Verd Islands) we spent three weeks. The geology was pre-eminently interesting, and I believe quite new: there are some facts on a large scale, of upraised coast that would interest Mr Lyell. . . . .
. . . .

St Jago is singularly barren, and produces few plants or insects; so that my hammer was my usual companion. . . . .
. . . .

On the coast I collected many marine animals, chiefly gasteropodous mollusca (I think some new). I examined pretty accurately a Caryophyllia, and, if my eyes were not bewitched, former descriptions have not the slightest resemblance to the animal. I took several specimens of an Octopus, which possessed a most marvellous power of changing its colours; equalling any chamelion, and evidently accommodating the changes to the colour of the ground which it passed over.   .   .   .   .   .

.   .   .   .

We then sailed for Bahia, and touched at the rock of St Paul. This is a serpentine formation.   .   .

.   .   .   .

After touching at the Abrothos, we arrived here on April 4th.   .   .   .   .

.   .   .   .

A few days after arriving, I started on an expedition of one hundred and fifty miles to Rio Macao, which lasted eighteen days.   .   .   .   .   .

.   .   .   .

I am now collecting fresh water and land animals: if what was told me in London is true, viz. that there are no small insects in the collections from the Tropics, I tell entomologists to look out and have their pens ready for describing. I have taken as minute (if not more so) as in England, Hydropori, Hygroti, Hydrobii, Pselaphi, Staphylini, Curculiones, Bembidia, &c. &c. It is exceedingly interesting to observe the difference of genera and species from those which I know; it is however much less than I had expected.   .   .   .   .   .   .

.   .   .   .

I have just returned from a walk, and as a specimen how little the insects are known, Noterus, according to Dic. Class. consists solely of three European species. I, in one haul of my net, took five distinct species.

At Bahia, the pegmatite and gneiss in beds had the same direction as was observed by Humboldt to prevail over Columbia, distant thirteen hundred miles.

MONTE VIDEO, *Aug.* 15, 1832.

My collection of plants from the Abrothos is interesting, as I suspect it contains nearly the whole flowering vegetation.

I made an enormous collection of Arachnidæ at Rio. Also a good many small beetles in pill-boxes: but it is not the best time of year for the latter.

Amongst the lower animals, nothing has so much interested me as finding two species of elegantly coloured planariæ (?) inhabiting the dry forest! The false relation they bear to snails is the most extraordinary thing of the kind I have ever seen. In the same genus (or more truly, family) some of the marine species possess an organization so mar-

vellous, that I can scarcely credit my eyesight. Every
one has heard of the discoloured streaks of water in
the equatorial regions. One I examined was owing to
the presence of such minute Oscillatoria, that in each
square inch of surface there must have been at least
one hundred thousand present. .   .   .   .

.   .   .   .

I might collect a far greater number of specimens
of invertebrate animals if I took up less time over
each: but I have come to the conclusion, that two
animals with their original colour and shape noted
down, will be more valuable to naturalists than six
with only dates and place.   .   .   .   .

.   .   .   .

At this present minute we are at anchor in the
mouth of the river: and such a strange scene it is.
Every thing is in flames—the sky with lightning—
the water with luminous particles—and even the very
masts are pointed with a blue flame.

Monte Video, *Nov.* 24, 1832.

We arrived here on the 24th of October, after
our first cruize on the coast of Patagonia, north of
the Rio Negro.   .   .   .   .   .   .

.   .   .   .

I had hoped for the credit of dame Nature, no such
country as this last existed; in sad reality we coasted
along two hundred and forty miles of sand hillocks;
I never knew before, what a horrid ugly object a
sand hillock is: the famed country of the Rio Plata
in my opinion is not much better; an enormous
brackish river bounded by an interminable green
plain, is enough to make any naturalist groan.

. . . .

I have been very lucky with fossil bones; I have
fragments of at least six distinct animals; as many
of these are teeth, shattered and rolled as they have
been, I trust they will be recognized. I have paid
all the attention I am capable of, to their geological
site; but of course it is too long a story for a
letter. 1st. the tarsi and meta-tarsi, very perfect, of
a cavia; 2d. the upper jaw and head of some very
large animal, with four square hollow molars, and the
head greatly produced in front. I at first thought
it belonged either to the megalonyx or megatherium.
In confirmation of this, in the same formation I found
a large surface of the osseous polygonal plates, which
"late observations" (what are they?) have shewn to
belong to the megatherium. Immediately I saw them
I thought they must belong to an enormous armadillo,
living species of which genus are so abundant here.
3d. The lower jaw of some large animal, which, from
the molar teeth I should think belonged to the
edentata: 4th. large molar teeth, which in some

respects would seem to belong to some enormous species of rodentia; 5th. also some smaller teeth belong-
ing to the same order, &c. &c.—They are mingled
with marine shells, which appear to me identical
with existing species. But since they were deposited
in their beds, several geological changes have taken
place in the country. . . . . .

. . . .

There is a poor specimen of a bird, which to my
unornithological eyes, appears to be a happy mixture of a lark, pigeon, and snipe. Mr Mac
Leay himself never imagined such an inosculating
creature. . . . .

. . . .

I have taken some interesting amphibia; a fine bipes;
a new Trigonocephalus, in its habits beautifully connecting Crotalus and Viperus: and plenty of new (as
far as my knowledge goes) saurians. As for one little
toad, I hope it may be new, that it may be christ-
ened "diabolicus." Milton must allude to this very
individual, when he talks of "squat like a toad."

. . . .

Amongst the pelagic crustaceæ, some new and curious
genera. Among Zoophites some interesting animals.
As for one Flustra, if I had not the specimen to back
me, nobody would believe in its most anomalous
structure. But as for novelty, all this is nothing
to a family of pelagic animals, which at first sight
appear like Medusa, but are really highly organized.

I have examined them repeatedly, and certainly from
their structure it would be impossible to place them
in any existing order. Perhaps Salpa is the nearest
animal; although the transparency of the body is
almost the only character they have in common.

. . . .

We have been at Buenos Ayres for a week. It is
a fine large city; but such a country; every thing
is mud; you can go no where, you can do nothing
for mud. In the city I obtained much information
about the banks of the Uruguay. I hear of limestone
with shells, and beds of shells in every direction.

. . . .

I purchased fragments of some enormous bones,
which I was assured belonged to the former giants!!

April 11, 1833.

We are now running up from the Falkland
Islands to the Rio Negro (or Colorado.) . .

. . . .

It is now some months since we have been at a
civilized port; nearly all this time has been spent
in the most southern part of Tierra del Fuego. It
is a detestable place; gales succeed gales at such
short intervals, that it is difficult to do any thing.
We were twenty-three days off Cape Horn, and could

by no means get to the westward.—We at last ran into harbour, and in the boats got to the west of the inland channels.—With two boats we went about three hundred miles; and thus I had an excellent opportunity of geologizing and seeing much of the savages. The Fuegians are in a more miserable state of barbarism than I had expected ever to have seen a human being. In this inclement country they are absolutely naked, and their temporary houses are like those which children make in summer with boughs of trees. . . .

. . . .

The climate in some respects is a curious mixture of severity and mildness; as far as regards the animal kingdom the former character prevails; I have in consequence not added much to my collections. The geology of this part of Tierra del Fuego was to me very interesting. The country is non-fossiliferous, and a common-place succession of granitic rocks and slates: attempting to make out the relation of cleavage, strata, &c. &c. was my chief amusement.

. . . .

The Southern ocean is nearly as sterile as the continent it washes. Crustaceæ have afforded me most work. . . . .

. . . .

I found a Zoea, of most curious form, its body being only one-sixth the length of the two spears. I am convinced, from its structure and other reasons, it is a young Erichthus. I must mention part of the structure of a Decapod, it is so very anomalous: the last pair of legs are small and dorsal, but instead of being terminated by a claw, as in all others, it has three curved bristle-like appendages; these are finely serrated and furnished with cups, somewhat resembling those of the Cephalopods. The animal being pelagic, this beautiful structure enables it to hold on to light floating objects. I have found out something about the propagation of that ambiguous tribe the Corallines. . . . .

. . . .

After leaving Tierra del Fuego, we sailed to the Falklands. . . . .

I had here the high good fortune to find amongst the most primitive looking rocks, a bed of micaceous sandstone, abounding with Terebratula and its subgenera, and Entrochites. As this is so remote a locality from Europe, I think the comparison of these impressions with those of the oldest fossiliferous rocks of Europe will be pre-eminently interesting. Of course they are only models and casts; but many of them are very perfect.

Rio de la Plata, *July* 18, 1833.

THE greater part of the winter has been passed in this river at Meldonado. . . . .

. . . .

We have got almost every bird in this neighbourhood, (Meldonado) about eighty in number, and nearly twenty quadrupeds. . . . . .

In a few days we go to the Rio Negro to survey some banks. . . . .

The geology must be very interesting. It is near the junction of the Megatherium and Patagonian cliffs. From what I saw. of the latter, in one half hour, in St Joseph's bay, they would be well worth a long examination. Above the great oyster-bed there is one of gravel, which fills up inequalities in its interior; and above this, and therefore high out of the water, is one of such modern shells that they retain their colour and emit a bad smell when burnt. Patagonia must clearly have lately risen from the water.

MONTE VIDEO, *November* 12, 1833.

I LEFT the Beagle at the Rio Negro, and crossed by land to Buenos Ayres. There is now carrying on a bloody war of extermination against the Indians, by which I was able to make this passage. But at the best it is sufficiently dangerous, and till now very rarely travelled. It is the most wild, dreary plain imaginable, without settled inhabitant or head of cattle. There are military "postas" at wide intervals, by which means I travelled. We lived for many days on deer and ostriches, and had to sleep in the open camp. . . . .

I had the satisfaction of ascending the Tierra de la Ventana, a chain of mountains between three and four thousand feet high, the very existence of which is scarcely known beyond the Rio Plata. After resting a week at Buenos Ayres, I started for St Fé. On the road the geology was interesting. I found two great groups of immense bones, but so very soft as to render it impossible to remove them. I think, from a fragment of one of the teeth, they belonged to the Mastodon. In the Rio Carcarana, I got a tooth which puzzles even my conjectures. It looks like an enormous gnawing one. At St Fé, not being well, I embarked and had a fine sail of three hundred miles down that princely river the Parana. When I returned to Buenos Ayres, I found the country upside down with revolutions, which caused me much trouble. I at last got away and joined the Beagle.

E. FALKLAND ISLAND, *March* 1834.

I HAVE been alarmed by your expression "cleaning all the bones," as I am afraid the printed numbers will be lost: the reason I am so anxious they should

not be, is, that a part were found in a gravel with recent shells, but others in a very different bed. Now with these latter there were bones of an Agouti, a genus of animals, I believe, peculiar to America, and it would be curious to prove that some one of the same genus coexisted with the megatherium; such, and many other points entirely depend on the numbers being carefully preserved.   .   .   .
.   .   .   .

I collected all the plants which were in flower on the coast of Patagonia, at Port Desire, and St Julian; also on the eastern parts of Tierra del Fuego, where the climate and features of Tierra del Fuego and Patagonia are united.   .   .   .   .   .
.   .   .   .

The soil of Patagonia is very dry, gravelly, and light. In East Tierra, it is gravelly, peaty, and damp. Since leaving the Rio Plata I have had some opportunities of examining the great southern Patagonian formation. I have a good many shells; from the little I know of the subject, it must be a tertiary formation, for some of the shells (and corallines) now exist in the sea. Others, I believe, do not. This bed, which is chiefly characterized by a great oyster, is covered by a very curious bed of porphyry pebbles, which I have traced for more than seven hundred miles. But the most curious fact is, that the whole of the east coast of the southern part of South America has been elevated from the ocean, since a period during which muscles have not lost their blue colour. At Port St Julian I found some very perfect bones of some large animal, I fancy a Mastodon: the bones of one hind extremity are very perfect and solid. This is interesting, as the latitude is between 49° and 50°, and the site far removed from the great Pampas, where bones of the narrow toothed Mastodon are so frequently found. By the way this Mastodon and the Megatherium, I have no doubt, were fellow brethren in the ancient plains. Relics of the Megatherium I have found at a distance of nearly six hundred miles in a north and south line. In Tierra del Fuego I have been interested in finding some sort of ammonite (also I believe found by Capt. King) in the slate near Port Famine; and on the eastern coast there are some curious alluvial plains, by which the existence of certain quadrupeds in the islands can clearly be accounted for. There is a sandstone with the impression of leaves like the common beech tree; also modern shells, &c., and on the surface of the table land there are, as usual, muscles with their blue colour, &c. . . . . .

. . . .

I have chiefly been employed in preparing myself for the South Sea, and examining the polypi of the smaller corallines in these latitudes. Many in themselves are very curious, and I think undescribed: there was one appalling one, allied to a Flustra, which I dare say I mentioned having found to the northward, where the cells have a moveable organ (like a vulture's head, with a dilatable beak), fixed on the edge. But what is of more general interest, is the unquestionable (as it appears to me) existence of another species of ostrich besides the Struthio ostrea. All the Gauchos and Indians state it is the case: and I place the greatest faith in their observations. I have the head, neck, piece of skin, feathers, and legs of one. The differences are chiefly in the colour of the feathers and scales; in the legs being feathered below the knees; also in its nidification, and geographical distribution.

VALPARAISO, *July* 24, 1834.

AFTER leaving the Falklands, we proceeded to the Rio Santa Cruz; followed up the river till within twenty miles of the Cordilleras: unfortunately want of provisions compelled us to return. This expedition was most important to me, as it was a transverse section of the great Patagonian formation. I conjecture (an accurate examination of the fossils may possibly determine the point) that the main bed is somewhere about the meiocene period (using Mr Lyell's expression); judging from what I have seen of the present shells of Patagonia. This bed contains an *enormous* mass of lava. This is of some interest, as being a rude approximation to the age of the volcanic part of the great range of the Andes. Long before this it existed as a slate and porphyritic line of hills. I have collected a tolerable quantity of information respecting the various periods and forms of elevations of these plains. I think these will be interesting to Mr Lyell. I had deferred reading his third volume till my return; you may guess how much pleasure it gave me; some of his wood-cuts came so exactly into play, that I have only to refer to them, instead of redrawing similar ones. .

. . . .

The valley of Santa Cruz appears to me a very curious one; at first it quite baffled me. I believe I can shew good reasons for supposing it to have been once a northern strait, like that of Magellan. . .

. . . .

In Tierra del Fuego I collected and examined some corallines: I have observed one fact which quite startled me. It is, that in the genus Sertularia (taken in its most restricted form as by Lamouroux), and in two species which, excluding comparative expressions, I should find much difficulty in describing as different, the polypi quite and essentially differed in all their most important and evident parts of structure. I have already seen enough to be convinced that the present families of corallines, as arranged by Lamarck, Cuvier, &c. are highly artificial. It appears to me, that they are in the same

B

state in which shells were, when Linnæus left them
for Cuvier to re-arrange.   .   .   .   .

.   .   .   .

It is most extraordinary I can no where see in my
books a single description of the polypus of any one
coral (excepting Lobularia (Alcyonium) of Savigny).
I found a curious little stony Cellaria (a new genus),
each cell provided with a long toothed bristle
capable of various and rapid motions. This motion
is often simultaneous, and can be produced by
irritation. This fact, as far as I see, is quite
isolated in the history (excepting of the Flustra, with
an organ like a vulture's head) of Zoophites. It
points out a much more intimate relation between
the polypi, than Lamarck is willing to allow. I forget
whether I mentioned having seen something of the
manner of propagation in that most ambiguous
family, the corallines: I feel pretty well convinced
that if they are not plants, they are not Zoophites:
the "gemmule" of a Halimeda contains several arti-
culations united, ready to burst their envelope and
become attached to some basis. I believe that in
Zoophites universally, the gemmule produces a single
polypus, which afterwards or at the same time grows
with its cell, or single articulation. The Beagle left
the strait of Magellan in the middle of winter; she
found her road out by a wild unfrequented channel;
well might Sir J. Nasborough call the west coast
South Desolation, "because it is so desolate a land

to behold." We were driven into Chiloe, by some very bad weather. An Englishman gave me three specimens of that very fine lucanoidal insect, which is described in the Cambridge Philosophical Transactions, two males and one female. I find Chiloe is composed of lava and recent deposits. The lavas are curious, from abounding with or rather being com-- posed of pitchstone.   .   .   .   .   .
.   .   .   .

We arrived here the day before yesterday; the views of the distant mountains are most sublime and the climate delightful; after our long cruise in the damp gloomy climates of the South, to breathe a clear dry air, and feel honest warm sunshine, and eat good fresh roast beef, must be the summum bonum of human life. I do not like the looks of the rocks half so much as the beef, there is too much of those rather insipid ingredients mica, quartz, and feldspar.
.   .   .   .

Shortly after arriving here I set out on a geological excursion, and had a very pleasant ramble about the base of the Andes. The whole country appears composed of breccias, (and I imagine slates) which universally have been modified, and often com- pletely · altered by the action of fire; the varieties of porphyry thus produced are endless, but no where have I yet met with rocks which have flowed in a stream; dykes of greenstone are very numerous. Modern volcanic action is entirely shut up in the very central parts (which cannot now be reached on account of the snow) of the Cordilleras. To the south of the Rio Maypo, I examined the tertiary plains already partially described by M. Gay. The fossil shells appear to me to differ more widely from the recent ones, than in the great Patagonian formation; it will be curious if an eocene and meiocene formation (recent there is abundance of) could be proved to exist in South America as well as in Europe. I have been much interested by finding abundance of recent shells at an elevation of thirteen hundred feet; the country in many places is scattered over with shells, but these are all *littoral* ones! So that I suppose the thirteen hundred feet elevation must be owing to a succession of small elevations, such as in 1822. With these certain proofs of the recent residence of the ocean over all the lower parts of Chili, the outline of every view and the form of each valley possesses a high interest. Has the action of running water or the sea formed this ravine? was a question which often arose in my mind, and was generally answered by my finding a bed of recent shells at the bottom. I have not sufficient arguments, but I do not believe that more than a small fraction of the height of the Andes has been formed within the tertiary period.

VALPARAISO, *March* 1835.

WE are now lying becalmed off Valparaiso, and I will take the opportunity of writing a few lines to you. The termination of our voyage is at last decided on. We leave the coast of America in the beginning of September, and hope to reach England in the same month of 1836.   .   .

.   .   .   .

You will have heard an account of the dreadful earthquake of the 20th of February. I wish some of the geologists, who think the earthquakes of these times' are trifling, could see the way in which the solid rock is shivered. In the town there is not one house habitable; the ruins remind me of the drawings of the desolated eastern cities. We were at Valdivia at the time, and felt the shock very severely. The sensation was like that of skating over very thin ice; that is, distinct undulations were perceptible. The whole scene of Concepcion and Talcuana is one of the most interesting spectacles we have beheld since leaving England. Since leaving Valparaiso, during this cruise, I have done little excepting in geology. In the modern tertiary strata I have examined four bands of disturbance, which reminded me on a small scale of the famous tract in the Isle of Wight. In one spot there were beautiful examples of three different forms of upheaval. In two cases I think I can show that the inclination is owing to the presence of a system of parallel dykes traversing the inferior mica slate. The whole of the coast from Chiloe to the south extreme of the Peninsula of Tres Montes is composed of the latter rock; it is traversed by very numerous dykes, the mineralogical nature of which will, I suspect, turn out very curious. I examined one grand transverse chain of granite, which has clearly burst up through the overlying slate. At the Peninsula of Tres Montes there has been an old volcanic focus, which corresponds to another in the north part of Chiloe. I was much pleased at Chiloe by finding a thick bed of recent oyster-shells, &c. capping the tertiary plain, out of which grew large forest trees. I can now prove that both sides of the Andes have risen in this recent period to a considerable height. Here the shells were three hundred and fifty feet above the sea. In Zoology I have done but very little; excepting a large collection of minute diptera and hymenoptera from Chiloe. I took in one day, Pselaphus, Anaspis, Latridius, Leiodes, Cercyon, and Elmis, and two beautiful true Carabi; I might almost have fancied myself collecting in England. A new and pretty genus of nudibranch mollusca which cannot crawl on a flat surface, and a genus in the family of balanidæ, which has not a true case, but lives in minute cavities in the shells of the concholapas, are nearly the only two novelties.

VALPARAISO, *April* 18, 1835.

I HAVE just returned from Mendoza, having crossed the Cordilleras by two passes. This trip has added much to my knowledge of the geology of the country. . . . .

. . . .

I will give a very short sketch of the structure of these huge mountains. In the Portillo pass (the more southern one) travellers have described the Cordilleras to consist of a double chain of nearly equal altitude, separated by a considerable interval. This is the case: and the same structure extends northward to Uspellata. The little elevation of the eastern line (here not more than six thousand or seven thousand feet) has caused it almost to be overlooked. To begin with the western and principal chain, where the sections are best seen; we have an enormous mass of a porphyritic conglomerate resting on granite. This latter rock seems to form the nucleus of the whole mass, and is seen in the deep lateral valleys, injected amongst, upheaving, overturning in the most extraordinary manner, the overlying strata. On the bare sides of the mountains, the complicated dykes and wedges of variously coloured rocks, are seen traversing in every possible form and shape the same formations, which, by their intersections, prove a succession of violences. The stratification in all the mountains is beautifully distinct, and owing to a variety in their colouring, can be seen at great distances. I cannot imagine any part of the world presenting a more extraordinary scene of the breaking up of the crust of the globe, than these central peaks of the Andes. The upheaval has taken place by a great number of (nearly) north and south lines*; which in most cases has formed as many anticlinal and synclinal ravines. The strata in the highest pinnacles are almost universally inclined at an angle from 70° to 80°. I cannot tell you how much I enjoyed some of these views; it is worth coming from England, once to feel such intense delight. At an elevation of from ten to twelve thousand feet, there is a transparency in the air, and a confusion of distances, and a sort of stillness, which give the sensation of being in another world; and when to this is joined the picture so plainly drawn of the great epochs of violence, it causes in the mind a most strange assemblage of ideas. The formation which I call porphyritic conglomerates, is the most important and most developed in Chili. From a great number of sections, I find it to be a true coarse conglomerate or breccia, which passes by every step, in slow gradation, into a fine clay-stone porphyry; the pebbles and cement becoming porphyritic, till at last all is blended in one compact rock. The porphyries are excessively abundant in this chain, and I feel sure that at least four-fifths of them have been thus

* Of dykes?

produced from sedimentary beds in situ. There are also porphyries which have been injected from below amongst the strata, and others ejected which have flowed in streams: and I could shew specimens of this rock, produced in these three methods, which cannot be distinguished. It is a great mistake to consider the Cordilleras (here) as composed only of rocks which have flowed in streams. In this range I no-where saw a fragment which I believe to have thus originated, although the road passes at no great distance from the active volcanos. The porphyries, conglomerates, sandstone, quartzone-sandstone, and limestones alternate and pass into each other many times (overlying clay-slate, when not broken through by the granite.) In the upper parts the sandstone begins to alternate with gypsum, till at last we have this substance of a stupendous thickness. I really think the formation is in some places (it varies much) nearly two thousand feet thick. It occurs often with a green (Epidote?) siliceous sandstone and snow-white marble : and resembles that found in the Alps, from containing large concretions of a crystalline marble of a blackish-gray colour. The upper beds, which form some of the higher pinnacles, consist of layers of snow-white gypsum and red compact sandstone, from the thickness of paper to a few feet, alternating in an endless round. The rock has a most curiously painted appearance. At the pass of the Puquenas in this formation, where a black rock (like clay-slate,

without many laminæ) and pale limestone have re-
placed the red sandstone, I found abundant impressions of shells. The elevation must be between twelve
thousand and thirteen thousand feet. A shell which
I believe is a Gryphæa is the most abundant. There
is also an Ostrea, Turritella, Ammonites, small bivalve,
Terebratula (?) Perhaps some good conchologist will
be able to give a guess to what grand division of the
continents of Europe these organic remains bear most
resemblance. They are exceedingly imperfect and few;
the Gryphites are most perfect. It was late in the
season, and the situation particularly dangerous, from
snow-storms. I did not dare to delay, otherwise a good
harvest might have been reaped. So much for the
western line. In the Portillo pass, proceeding east-
ward, I met with an immense mass of a conglomerate,
dipping to the west 45°, which rests on micaceous
sandstone, &c. upheaved, converted into quartz rock,
penetrated by dykes, from a very grand mass of
protogene (large crystals of quartz, red feldspar,
and a little chlorite.) Now this conglomerate, which
reposes on and dips from the protogene at an
angle of 45°, consists of the peculiar rocks of the
first described chain, *pebbles* of the black rock
with shells, green sandstone, &c. &c. It is here
manifest also that the upheaval (and deposition at
least of part) of the grand eastern chain is entirely
posterior to the western. To the north, in the
Uspellata pass, we have also a fact of the same

class. Bear this in mind; it will help to make you
believe what follows. I have said the Uspellata range
is geologically, although only six thousand or seven
thousand feet, a continuation of the grand eastern
chain. It has its nucleus of granite, consisting of
grand beds of various crystalline rocks, which I can
feel no doubt are subaqueous lavas alternating with
sandstone, conglomerates, and white aluminous beds
(like decomposed feldspar) with many other curious
varieties of sedimentary deposits. These lavas and
sandstones alternate very many times, and are quite
conformable one to the other. During two days of
careful examination I said to myself at least fifty
times, how exactly like, only rather harder, these
beds are to those of the upper tertiary strata of
Patagonia, Chiloe, and Concepcion, without the possibility of their identity ever having occurred to me.
At last there was no resisting the conclusion. I could
not expect shells, for they never occur in this formation; but lignite or carbonaceous shale ought to be
found. I had previously been exceedingly puzzled by
meeting in the sandstone with thin layers (a few inches
to some feet thick) of a brecciated pitchstone. I now
strongly suspect that the underlying granite has altered
such beds into this pitchstone. The silicified wood
(particularly characteristic of the formation) was yet
absent; but the conviction that I was on the tertiary
strata was so strong in my mind by this time, that on
the third day, in the midst of lavas, and heaps of

granite, I began an apparently forlorn hunt in search
of it. How do you think I succeeded? In an escarpe-
ment of compact greenish sandstone I found a small
wood of petrified trees in a vertical position, or rather
the strata were inclined about 20° or 30° to one point,
and the trees 70° to the opposite; that is, they were
before the tilt truly vertical. The sandstone consists of many horizontal layers, and is marked by
the concentric lines of the bark (I have a specimen).
Eleven are perfectly silicified, and resemble the dicotyledonous wood which I found at Chiloe and Concepcion: the others, thirty to thirty-four in number,
I only know to be trees from the analogy of form
and position; they consist of snow-white columns (like
Lot's wife) of coarsely crystalized carbonate of lime.
The largest shaft is seven feet. They are all close
together, within one hundred yards, and about the
same level; no where else could I find any. It
cannot be doubted that the layers of fine sandstone
have quietly been deposited between a clump of trees,
which were fixed by their roots. The sandstone rests
on lava, is covered by a great bed, apparently about
one thousand feet thick, of black augitic lava, and over
this there are at least five grand alternations of such
rocks and aqueous sedimentary deposits; amounting in
thickness to several thousand feet. I am quite afraid
of the only conclusion which I can draw from this
fact, namely, that there must have been a depression
in the surface of the land to that amount. But

neglecting this consideration, it was a most satis-
factory support of my presumption of the tertiary
age of this eastern chain. (I mean by tertiary, that
the shells of the period were closely allied to, and
some identical with, those which now lie in the lower
beds of Patagonia.) A great part of the proof must
remain upon my ipse dixit of a mineralogical resemblance to those beds whose age is known. According
to this view granite, which forms peaks of a height
probably of fourteen thousand feet, has been fluid in
the tertiary period: strata of that period have been
altered by its heat, and are traversed by dykes from
the mass: are now inclined at high angles, and form
regular or complicated anticlinal lines. To complete
this climax, these same sedimentary strata and lavas
are traversed by very numerous true metallic veins of
iron, copper, arsenic, silver, and gold, and these can
be traced to the underlying granite. A gold mine
has been worked close to the clump of silicified trees.
When you see my specimens, sections, and account,
you will think there is pretty strong presumptive evi-
dence of the above facts. They appear very important;
for the structure and size of this chain will bear comparison with any in the world: and that all this
should have been produced in so very recent a period
is indeed remarkable. In my own mind I am quite
convinced of it. I can any how most conscientiously
say, that no previously formed conjecture warped my
judgment. As I have described, so did I actually

observe the facts. .   .   .   .   .   .

On some of the large patches of perpetual snow, I found the famous red snow of the arctic regions. I send with this letter my observations and a piece of paper on which I tried to dry some specimens.

I also send a small bottle with two Lizards: one of them is viviparous, as you will see by the accompanying notice. M. Gay, a French naturalist, has already published in one of the newspapers of this country a similar statement, and probably has forwarded to Paris some account*. .   .

---

* The following is an Extract from the Newspaper referred to by Mr DARWIN:

"Besides these labours I employed myself during the great rains in dissecting various reptiles. It must be interesting to know the influence of the climate of Valdivia on the animals of this family. In the greater part of those which I have been able to submit to my scalpel, I have found a truly extraordinary fact, that they were viviparous. Not only the innocent Snake of Valdivia has offered to my notice this singular phenomenon, but also a beautiful and new kind of Iguana which approaches very near to the *Liposoma* of Spix, and to which, on account of its beautiful colours, he has given the name of *Chrysosaurus*. All the species, even those which lay eggs in Santiago, here produce their young alive; and the same thing happens with some *Batrachians*, and particularly with a genus near to the *Rhinella* of Fitzingen, of which the numerous species have the skin pleasingly spotted with green, yellow, and black. I need not dwell on the importance of this last example, in reference to comparative anatomy: an importance which appeared to me still greater when, on analyzing a Tadpole not yet transformed, I satisfied myself that nature has not varied

. . . .

In the box there are two bags of seeds, one ticketed
" valleys of Cordilleras five thousand to ten thousand
feet high"; the soil and climate exceedingly dry; soil
light and strong; extremes in temperature: the other
" chiefly from the dry sandy traversia of Mendoza, three
thousand feet, more or less." If some of the bushes
should grow, but not be healthy, try a slight sprink-
ling of salt and saltpetre. The plain is saliferous.

. . . .

In the Mendoza bag, there are the seeds or berries
of what appears to be a small potatoe plant with a
whitish flower. They grow many leagues from where
any habitation could ever have existed, owing to the
absence of water. Amongst the Chonos dried plants,
you will see a fine specimen of the wild potatoe,
growing under a most opposite climate, and unques-
tionably a true wild potatoe. It must be a distinct
species from that of the lower Cordilleras.

her plan of organization. In these, as in the Tadpoles which
live in water, the intestines were of a length very dispropor-
tioned to the body: now if this length was necessary to the latter,
which live upon vegetable substances, it was altogether useless to
those which are to undergo their metamorphosis in the belly
of the mother: and thus nature has followed the march pre-
scribed to her by a uniformity of construction, and without
deviating from it, has admitted a simple exception, a real hiatus,
well worthy the attention of the philosophical naturalist."

ADDRESS

TO THE

Members of the University of Cambridge,

ON THE EXPEDIENCY OF IMPROVING, AND ON THE FUNDS

REQUIRED FOR REMODELLING AND SUPPORTING,

THE BOTANIC GARDEN.

BY THE REV. J. S. HENSLOW, M.A.,

ST. JOHN'S COLLEGE, AND PROFESSOR OF BOTANY IN THE UNIVERSITY OF CAMBRIDGE.

CAMBRIDGE:

PRINTED BY METCALFE AND PALMER.

1846.

Price One Shilling.

AN ADDRESS,

ALTHOUGH the Grace recommended to the Senate
by the Botanic Garden Syndicate has been rejected,
I understand the minority consisted of more than
two-fifths of the members who voted on the oc-
casion. From this circumstance, and from what
I know of the opinion of some of the influential
Members of the Senate, I presume the rejection of
this Grace (which merely went to the extent of
proposing another Syndicate) must be taken rather
as a decided expression of opinion against the expediency of laying any tax whatever upon the
University for the object proposed, than as a deter-
mination to resist all improvement in an establishment which has become utterly unsuited to the
demands of modern science. I shall therefore ven-
ture to address a few observations to the members
of our University, inviting attention to what may
be considered requisite for a modern Botanic Garden ; and to enquire whether some means cannot
be devised for obviating the necessity of our taking
any step which might be viewed as a retrogade

movement, so far as respects the interests of Botany among us. But before I do this, I trust a few observations concerning the position which Botany and the other Natural Sciences occupy in Cambridge, will not be considered as an attempt on my part to cast any slur upon the resident members of the University, for their not having done what they scarcely possess the power of doing, without either a direct tax, or diverting funds which have been entrusted to them for specific purposes, to objects for which they were not originally destined to be applied. I claim to be as warm an admirer of so much of that sound learning and religious education as is upheld by us, as any other son of our Alma Mater. No one who has spent the same number of happy years within the precincts of our University as I have done, or who has experienced the same amount of kindness and friendship there, could ever wilfully cast a slur upon its residents without being grossly ungrateful for past enjoyment, and wholly unfitted for any that may yet be in store for him. But it is the very desire to see our University improving, that induces me to state what I believe to be a few plain facts, and to give expression to the opinions of one who for many years took no inactive part in the encouragement of Natural History to the little extent which his opportunities allowed him.

I hope I should be among the last to reflect upon the resident members of the Senate for any

supposed backwardness about upholding the particular science for whose encouragement I am more
especially bound to plead. I most fully and grate-
fully bear my testimony, that from the time I first
occupied the Botanical Chair, I have ever found the
Senate ready to accede to whatever proposal I have
made to them, from time to time, for rendering me
assistance, either in purchasing specimens for our
Botanical Museum, or in providing the means for
securing their preservation. But still I must consider the claims of Botany are not sufficiently ap-
preciated among us. There are persons of great
mathematical and classical attainments who have
very erroneous notions respecting the ultimate aim
and object of this science. Many persons, both
within and without the Universities, suppose its
objects limited to fixing names to a vast number
of plants, and to describing and classing them
under this or that particular " system." They are
not aware that systematic Botany is now considered
to be no more than a necessary stepping-stone to
far more important departments of this science,
which treat of questions of the utmost interest to
the progress of human knowledge in certain other
sciences which have been more generally admitted
to be essential to the well-being of mankind. For
instance, the most abstruse speculations on animal
physiology are to be checked, enlarged, and guided
by the study of vegetable physiology. Without
continued advances in this latter department of

Botany, the progress towards perfection in general physiology must be comparatively slow and uncertain. As regards the progress of Botanical physiology, even Chemistry itself must be viewed as a subordinate assistant, whilst it is making us acquainted with those physical forces by which mere brute matter is regulated and arranged. Those forces are themselves to be restrained and modified by the instrumentality of vegetable life, in bodies whose appointed position is to prepare all the organic matter that is destined for the support of a still higher race of creatures in the general scheme of nature. We may feel quite confident that some of the arts we consider to be most important to man, such as Agriculture and Horticulture, will never be perfected until the fundamental principles of vegetable physiology shall have been satisfactorily elucidated. Numerous indeed are the bearings, direct and indirect, which Botany holds upon other sciences, and upon various arts! And, here, I cannot refrain from making some allusion to the station which Botany might be made to occupy in an improved scheme of liberal education. But as this claim has been so fully recognized by Dr. Whewell, in his late publication " On Cambridge Studies," I shall prefer reminding you of his opinion, to urging upon you whatever may be my own view.

" I have said that a portion of the sciences which " have come into existence in modern times, and

" which are still in progress, should be introduced
" into a liberal education, to such an extent as to
" acquaint the student with their nature and prin-
" ciples. It is an important enquiry, in determin-
" ing the proper scheme of a liberal education,
" what portion of science is best fitted for this pur-
" pose. I have already remarked elsewhere, that
" among the sciences, Natural History affords very
" valuable lessons which may beneficially be made
" a portion of education : the more so, inasmuch as
" this study may serve to correct prejudices and
" mental habits which have often been cherished by
" making pure mathematics the main instrument
" of intellectual education. The study of Natural
" History teaches the student that there may be an
" exact use of names, and an accumulated store of
" indisputable truths, in a subject in which names
" are not appropriated by definitions, but by the
" condition that they shall serve for the expression
" of truth. These sciences show also that there
" may exist a system of descriptive terms which
" shall convey a conception of objects almost as
" distinct as the senses themselves can acquire for
" us, at least when the senses have been educated
" to respond to such a terminology. Botany, in
" particular, is a beautiful and almost perfect ex-
" ample of these scientific merits; and an acquaint-
" ance with the philosophy of Botany will supply
" the student with a portion of the philosophy of
" the progressive sciences, highly important, but for

" the most part hitherto omitted in the usual plans
" of a liberal education. But the philosophy of
" Botany cannot be really understood without an
" acquaintance with a considerable portion, at least,
" of the details of systematic Botany. On these
" grounds, I should much desire to see Botany, or
" some other branch of Natural History, or Natural
" History in general, introduced as a common
" element into our higher education, and recom-
" mended to the study of those who desire to have
" any clear view of the nature of the progressive
" sciences; since it is, in fact, the key and ground-
" work of a large portion of those sciences."

Very possibly, as some suppose, the time may
not yet have arrived for these sentiments to meet
with any loud response among us; and certainly I
am not to be considered a competent judge on this
point. Still I trust you will allow me to remind
you that we are very often blamed by persons who
do not understand the peculiarities of our establishment, for not doing more than we at present attempt
for the natural sciences. All know that to a limited
extent some of them are encouraged; but they
know also that not one of them is patronized,
like Mathematics and Classics, by the stimulus of
rewards for securing its diligent and successful
cultivation. What Dr. Whewell has said at p. 224,
concerning the establishment of a General Tripos
for students in the progressive sciences, appears to
me well worthy of our most serious attention.

Every one who has had much experience in preparing pupils for an Ordinary Degree can bear testimony to the fact of there being minds naturally incapacitated for clearly and fully comprehending a mathematical problem. Now there are many persons with this want of mathematical ability, who will delightedly occupy themselves in one or other department of the natural sciences. It has long appeared to me a mistaken policy to bind down such persons to one particular routine of dull anxious plodding, without allowing them the opportunity, before quitting the University, of proving that they have not laboured unsuccessfully in the general field of human knowledge. But I shall not dwell upon the various and numerous considerations which incline me to accept the view that has been already taken by Dr. Whewell, of the propriety of our adding Botany, or some other branch of Natural History, to the general curriculum of a liberal education.

There are other grounds for enforcing my appeal to you, which seem to me more likely to awaken your sympathies in favour of the scheme that has been proposed for extending our Garden. Surely a long-established University like ours ought, if possible, so far to keep pace with the advance of every science, as to give her resident members the greatest facility for cultivating and improving any one branch in particular. Even if Botany should never be recognized by us as a description of learning that can be considered well qualified for "mental training," yet (I presume) no one who has any just conception of the Natural Sciences will deny that they are most worthy the attention of those who are already "mentally trained;" and consequently that they are no where more capable of being duly upheld than by many of the resident members of our University. That Botany does not flourish among us, and that this, in common with other progressive sciences, has greatly languished of late years, is too notorious to need further comment from me. When I shall presently state what I consider to be the defect of our system with respect to the Natural Sciences, it will not be from any desire to justify myself: but I hope to state the truth, and then leave others to make their own inferences as to whether sufficient improvements may not be made in our establishment for securing the continued progression of Botany, as well as of every useful and important branch of human learning, among us.

I shall invite attention to one further plea for our efficiently securing the progress of the Natural Sciences among us. I trust I am not too presumptuous in naming it to a University whose constantly avowed profession it is to connect "sound learning with religious education." Our appointed calling to "replenish the earth and subdue it," means something more than to exercise authority over the beasts of the field, or to prevent them

intruding upon any of the haunts of man. Has
it not been chiefly owing to the progress of such
sciences as have unravelled to us some portion of
the physical laws by which the works of the creation
have been advanced to their present condition, that
so many of those cobwebs of bye-gone supersti-
tions have been successfully brushed aside, which
once entangled the best exertions of some of the
foremost and most devoted champions of the truth
in former generations? Does not a partial disbelief
in some of the best established truths which our
improved acquaintance with God's works has at
length revealed to us, even now prevail with certain minds deeply imbued with scriptural learning,
to the extent of allowing them to put some trust
in means that are even beneath the admitted insufficiency of the beggarly elements of the law?
With whatever deference we may be prepared to
look up to men who are deeply learned in verbal
criticism, it is quite impossible for some of us (and
I expect for many of us) ever to believe that such
persons can have caught the spirit of the scriptures,
who will persist in couching their doctrines under
forms, and in clothing their discipline with obser-
vances, which are ill adapted to encourage an intelli-
gent people to advance in the observance of pure
and undefiled religion. Not many of our people are
likely to consent, I trust, to be again encumbered
with the idle fancies and vain delusions which a few
centuries back so much disfigured the worship of

their forefathers, until they were taught to cast aside these fancies as something worse than inexpedient. May they never be prevailed upon to despise those who have been the best expounders and defenders of the tenets of the pure and reformed section of the Church that is established in this kingdom! No one, I hope, will imagine that I have any idea that the resident members of our University are likely to advise that we should abandon the better for the worse. But we have seen that deep scriptural learning is insufficient to protect some minds against so fatal a mistake: and it may be well worthy our consideration whether a University can safely cease from esteeming it a duty to encourage any department of sound learning which may assist in improving our general views of God's providence. Whatever might have been excusable in ages of comparative ignorance, surely our present wisdom is to accept of every improved light we have now obtained from the works of God, as well as from every learnedly corrected interpretation of His Word: so that these two witnesses to His infinite power, wisdom and goodness, may mutually corroborate each other. Whilst our faith shall be set forth to us untainted by superstition, we shall be better able to keep our consciences as void of offence towards God as, by the practice of sound morals, we may contrive to do towards man. Where then ought all unworthy notions about the results to which the Natural

Sciences may yet conduct us, to meet with more decided contradiction and rebuke, than within the walls of a University? and yet how is this likely to happen again, if the tendency to neglect these sciences (I do not say to banish them) shall continue to increase, as most certainly it has of late years been doing among us?

If, then, it should appear to you sufficiently clear that these sciences ought to be much better supported and encouraged in this University than they are at present, or have ever been, I am thoroughly persuaded that some means might be devised for inviting and securing ample assistance for carrying out your wishes. Only let our wants be known in quarters whence efficient assistance can be easily given, and past experience should satisfy us that we have no right to despair about receiving it. Let me now bring before you an account of what seems to be required for securing success to the scheme that has been suggested for our New Botanic Garden.

At the request of the Trustees, I took some pains in assisting them last year to secure the services of a Curator, who should be competent to meet the demands which such an establishment as we hope to see among us may require. At their request also, and that of the Syndicate, I twice visited the National Establishment at Kew, and obtained from its talented and experienced superintendent, Sir W. Hooker, as well as from Dr. Lindley, and some others thoroughly acquainted with the details of Botanic Gardens, such information and advice as were calculated to put our own Garden upon at least an equal footing with those of Edinburgh, Glasgow, or Dublin. It appeared we should not be able to execute such a plan as we contemplated, and also to provide for its efficient support, unless the Trustees could command about ten thousand pounds more than they are likely to raise by disposing of the old site. Unless an adequate sum can be secured for supporting in cultivation, as well as laying out between twenty and thirty acres, we shall run the risk of following the example of some provincial establishments, which have flourished for a while under the excitement of novelty, and through the zeal of a few active supporters, but which have ultimately fallen into neglect, or even been sold to pay off debts that could not be met under the diminished amount of annual subscriptions.

Periods of no progress, or even of retrogression, may again arrive, like that which has afflicted our present Garden ; but still, the permanency of our establishment is secured to us by the nature of the foundation, and must continue (whilst the rights of property are respected) to form a nucleus around which fresh gatherings of strength may take place, and whence renewed vigour may emanate sufficient to raise it again to a level with some other establishments of the same kind, which are well calculated to keep the Botanists of our nation in advance of those of other countries, by the superior facilities which are thus afforded them. It is indeed true enough, that one man with half-a-dozen flower-pots may do more towards advancing Botany than another will feel inclined to attempt with twenty or thirty acres of garden at his command: but it may very safely be asserted, that the larger the number of living species that are cultivated in a Botanic Garden, the greater will be the facilities afforded us all; not merely for systematic improvement, but for anatomical and other experimental researches essential to the progress of general physiology. It is impossible to predict what particular species may safely be dispensed with in such establishments, without risking some loss of opportunity which that very species might have offered to a competent investigator, at the exact moment he most needed it. The reason why a modern Botanic Garden requires so much larger space than formerly, is chiefly owing to the vastly increased number of trees and shrubs that have been introduced within the last half century. The demands of modern science require as much attention to be paid to these, as to those herbaceous species which alone can form the staple of the collections in small establishments. The considerable portion of the ground which would be devoted to an Arboretum may be kept up at very much less expence than the rest, but would add very greatly to the ornamental as well as to the efficient character of the Garden.

I shall here take a slight review of what was the position of the Professor of Botany whilst I resided in the University. I do this, not with any desire of justifying my own insufficiency for advancing the science, or of exculpating myself from any merited rebuke for having neglected whatever facilities I possessed for so doing : whether they were offered me by the present Garden, or by many valuable works on Botany in our Public Library. But I do this in order that the public may be able to understand more thoroughly than they do at present, what are the peculiar disadvantages under which Botany and other Natural Sciences are to be upheld in our University, as well as the peculiar advantages we there enjoy for promoting them. I possess that sort of love for Natural History which makes the pursuit of any branch of it a delightful occupation to me. When appointed to the Botanical Chair I knew very little indeed about Botany, my attention having before that been devoted chiefly to other departments; though I probably knew as much of the subject as any other resident in Cambridge. There had been no lectures delivered in Botany during the last thirty years of the protracted life of my predecessor, who had not resided for several years before I came to Cambridge, and whom I never saw. After entering on my office I was deprived of the opportunity of paying an undivided attention to its duties, by the necessity under which I then lay of improving my income beyond the very scanty stipend which is attached to our Professorships. Five or six hours a-day devoted to cramming men for their degrees, is so far apt to weary the mind as to indispose it for laborious study: especially if we happen not to be gifted with talents or energy sufficient to overcome such an obstacle. Attention to the mere arrangement of an extensive herbarium also takes up a vast deal more time than persons not acquainted with the subject are at all aware of. Botany is now so very extensive a subject, that it needs the devoted study of many years to enable any one to master what has been done in it by others; and it requires the continued attention of a whole life to qualify a man for taking the lead in any one of its departments. Such a general knowledge, indeed, of Botany, as may qualify us for reaping satisfaction and delight from this study, may be readily attained by moderate diligence; and I have often received assurances from persons who have acquired a taste for Botany in our class-room, that they have found the pursuit a great enjoyment to them; especially when they have happened to be located in country districts where there was no one possessing sufficiently intellectual resources to be companionable. We can, no doubt, refer to periods when one or other of the sciences of Natural History have been vigorously promoted among us, owing to the superior advantages for self-devotion, or of mental activity, which some gifted individual may have possessed. Such an one

is able, in spite of all disheartening or discouraging
concomitants, to shed lustre upon the University,
by the foremost position he takes up in the front
of all Europe, as a leader in the particular science
he may have embraced.   But it is not fair, at least
it is not wise, to calculate by what has been done
at Cambridge in these respects, as to what might
still be done without further improvement upon our
present resources.   There is not sufficient provision
made for securing steadiness in the progression, or
even in the diffusion, of the Natural Sciences among
us.   It will not be supposed by you that I am anti-
cipating the possibility of improving my own salary
by advising an increase to the resources necessary for
conducting our Botanic Garden.   The Professor of
Botany has no power or authority over the Garden.
This is a private endowment, under the care of
a Curator, who is appointed by five Trustees, four
of whom are Heads of Colleges, and the fifth the
Regius Professor of Physic.   The Professorship of
Botany is a Government appointment, not connected
with the Garden establishment, otherwise than by the
Professor also enjoying the office of Walker's Lecturer
(without salary), he becomes entitled to make use of
it.   I therefore speak more freely to the wants of this
establishment than if I were more directly implicated
in its management.   I was not myself inclined to
suppose that it would be considered expedient to
tax the University as a body, for the improvement of
the Garden.   All persons who have long resided in

the University are well aware that óur resources
for promoting any general purpose that provided
for by special endowment, are of the most scanty
description. Indeed, doubts are frequently expressed, about the period when the annual audit
of our accounts takes place, whether we are solvent
or not. The world in general does not understand
the distinction that is to be made between those
resources which are at the command of the University, and those which are at the command of the
separate Colleges. Now Botany at present forms
no part of the appointed curriculum for our students, excepting *conditionally*, whilst the Professor
may continue to deliver twenty lectures; in which
case it is required that the medical students (about
half-a-dozen annually) should attend them. It
was not to be expected, therefore, that a majority
of the resident members should be disposed to tax
the whole University, in order to uphold an object
which forms so trivial a portion of our system,
however desirous they might be to see the Garden
enlarged for the more general advantage of this
science. I am persuaded that the University at
large will regret to find us compelled to abandon
a scheme, for the execution of which we may never
again possess such an opportunity as we at present
enjoy. The only method, then, which now presents
itself as at all likely to meet with success, is for us
to attempt raising the requisite sum by general sub-
scription among all who are members of the Univer-

sity, and among all who respect our institution. But if an effort of this kind is to be made, let it be one that is worthy of us; or else, I think, we had better leave matters as they are. I would therefore suggest that our subscriptions should be conditional, and not to be called for unless their sum total shall amount to a certain specified sum, which it will be no disgrace to us to appropriate to the advancement of science.

HITCHAM, *April* 4, 1846.

Printed in the United States
By Bookmasters